Lecture Notes in Mathematics

Edited by A. Dold and B. Eckmann

796

Corneliu Constantinescu

Duality in Measure Theory

Springer-Verlag
Berlin Heidelberg New York

Lecture Notes in Mathematics

For information about Vols. 1–580, please contact your bookseller or Springer-Verlag.

Vol. 581: Séminaire de Probabilités XI, Université de Strasbourg. Proceedings 1975/1976. Edité par C. Dellacherie, P. A. Meyer et M. Weil. VI, 574 pages. 1977.

Vol. 582: J. M. G. Fell, Induced Representations and Banach *-Algebraic Bundles. IV, 349 pages. 1977.

Vol. 583: W. Hirsch, C. C. Pugh and M. Shub, Invariant Manifolds. IV, 149 pages. 1977.

Vol. 584: C. Brezinski, Accélération de la Convergence en Analyse Numérique. IV, 313 pages. 1977.

Vol. 585: T. A. Springer, Invariant Theory. VI, 112 pages. 1977.

Vol. 586: Séminaire d'Algèbre Paul Dubreil, Paris 1975–1976 (29ème Année). Edited by M. P. Malliavin. VI, 188 pages. 1977.

Vol. 587: Non-Commutative Harmonic Analysis. Proceedings 1976. Edited by J. Carmona and M. Vergne. IV, 240 pages. 1977.

Vol. 588: P. Molino, Théorie des G-Structures: Le Problème d'Equivalence. VI, 163 pages. 1977.

Vol. 589: Cohomologie l-adique et Fonctions L. Séminaire de Géométrie Algébrique du Bois-Marie 1965–66, SGA 5. Edité par L. Illusie. XII, 484 pages. 1977.

Vol. 590: H. Matsumoto, Analyse Harmonique dans les Systèmes de Tits Bornologiques de Type Affine. IV, 219 pages. 1977.

Vol. 591: G. A. Anderson, Surgery with Coefficients. VIII, 157 pages. 1977.

Vol. 592: D. Voigt, Induzierte Darstellungen in der Theorie der endlichen, algebraischen Gruppen. V, 413 Seiten. 1977.

Vol. 593: K. Barbey and H. König, Abstract Analytic Function Theory and Hardy Algebras. VIII, 260 pages. 1977.

Vol. 594: Singular Perturbations and Boundary Layer Theory, Lyon 1976. Edited by C. M. Brauner, B. Gay, and J. Mathieu. VIII, 539 pages. 1977.

Vol. 595: W. Hazod, Stetige Faltungshalbgruppen von Wahrscheinlichkeitsmaßen und erzeugende Distributionen. XIII, 157 Seiten. 1977.

Vol. 596: K. Deimling, Ordinary Differential Equations in Banach Spaces. VI, 137 pages. 1977.

Vol. 597: Geometry and Topology, Rio de Janeiro, July 1976. Proceedings. Edited by J. Palis and M. do Carmo. VI, 866 pages. 1977.

Vol. 598: J. Hoffmann-Jørgensen, T. M. Liggett et J. Neveu, Ecole d'Eté de Probabilités de Saint-Flour VI – 1976. Edité par P.-L. Hennequin. XII, 447 pages. 1977.

Vol. 599: Complex Analysis, Kentucky 1976. Proceedings. Edited by J. D. Buckholtz and T. J. Suffridge. X, 159 pages. 1977.

Vol. 600: W. Stoll, Value Distribution on Parabolic Spaces. VIII, 216 pages. 1977.

Vol. 601: Modular Functions of one Variable V, Bonn 1976. Proceedings. Edited by J.-P. Serre and D. B. Zagier. VI, 294 pages. 1977.

Vol. 602: J. P. Brezin, Harmonic Analysis on Compact Solvmanifolds. VIII, 179 pages. 1977.

Vol. 603: B. Moishezon, Complex Surfaces and Connected Sums of Complex Projective Planes. IV, 234 pages. 1977.

Vol. 604: Banach Spaces of Analytic Functions, Kent, Ohio 1976. Proceedings. Edited by J. Baker, C. Cleaver and Joseph Diestel. VI, 141 pages. 1977.

Vol. 605: Sario et al., Classification Theory of Riemannian Manifolds. XX, 498 pages. 1977.

Vol. 606: Mathematical Aspects of Finite Element Methods. Proceedings 1975. Edited by I. Galligani and E. Magenes. VI, 362 pages. 1977.

Vol. 607: M. Métivier, Reelle und Vektorwertige Quasimartingale und die Theorie der Stochastischen Integration. X, 310 Seiten. 1977.

Vol. 608: Bigard et al., Groupes et Anneaux Réticulés. XIV, 334 pages. 1977.

Vol. 609: General Topology and Its Relations to Modern Analysis and Algebra IV. Proceedings 1976. Edited by J. Novák. XVIII, 225 pages. 1977.

Vol. 610: G. Jensen, Higher Order Contact of Submanifolds of Homogeneous Spaces. XII, 154 pages. 1977.

Vol. 611: M. Makkai and G. E. Reyes, First Order Categorical Logic. VIII, 301 pages. 1977.

Vol. 612: E. M. Kleinberg, Infinitary Combinatorics and the Axiom of Determinateness. VIII, 150 pages. 1977.

Vol. 613: E. Behrends et al., L^P-Structure in Real Banach Spaces. X, 108 pages. 1977.

Vol. 614: H. Yanagihara, Theory of Hopf Algebras Attached to Group Schemes. VIII, 308 pages. 1977.

Vol. 615: Turbulence Seminar, Proceedings 1976/77. Edited by P. Bernard and T. Ratiu. VI, 155 pages. 1977.

Vol. 616: Abelian Group Theory, 2nd New Mexico State University Conference, 1976. Proceedings. Edited by D. Arnold, R. Hunter and E. Walker. X, 423 pages. 1977.

Vol. 617: K. J. Devlin, The Axiom of Constructibility: A Guide for the Mathematician. VIII, 96 pages. 1977.

Vol. 618: I. I. Hirschman, Jr. and D. E. Hughes, Extreme Eigen Values of Toeplitz Operators. VI, 145 pages. 1977.

Vol. 619: Set Theory and Hierarchy Theory V, Bierutowice 1976. Edited by A. Lachlan, M. Srebrny, and A. Zarach. VIII, 358 pages. 1977.

Vol. 620: H. Popp, Moduli Theory and Classification Theory of Algebraic Varieties. VIII, 189 pages. 1977.

Vol. 621: Kauffman et al., The Deficiency Index Problem. VI, 112 pages. 1977.

Vol. 622: Combinatorial Mathematics V, Melbourne 1976. Proceedings. Edited by C. Little. VIII, 213 pages. 1977.

Vol. 623: I. Erdelyi and R. Lange, Spectral Decompositions on Banach Spaces. VIII, 122 pages. 1977.

Vol. 624: Y. Guivarc'h et al., Marches Aléatoires sur les Groupes de Lie. VIII, 292 pages. 1977.

Vol. 625: J. P. Alexander et al., Odd Order Group Actions and Witt Classification of Innerproducts. IV, 202 pages. 1977.

Vol. 626: Number Theory Day, New York 1976. Proceedings. Edited by M. B. Nathanson. VI, 241 pages. 1977.

Vol. 627: Modular Functions of One Variable VI, Bonn 1976. Proceedings. Edited by J.-P. Serre and D. B. Zagier. VI, 339 pages. 1977.

Vol. 628: H. J. Baues, Obstruction Theory on the Homotopy Classification of Maps. XII, 387 pages. 1977.

Vol. 629: W. A. Coppel, Dichotomies in Stability Theory. VI, 98 pages. 1978.

Vol. 630: Numerical Analysis, Proceedings, Biennial Conference, Dundee 1977. Edited by G. A. Watson. XII, 199 pages. 1978.

Vol. 631: Numerical Treatment of Differential Equations. Proceedings 1976. Edited by R. Bulirsch, R. D. Grigorieff, and J. Schröder. X, 219 pages. 1978.

Vol. 632: J.-F. Boutot, Schéma de Picard Local. X, 165 pages. 1978.

Vol. 633: N. R. Coleff and M. E. Herrera, Les Courants Résiduels Associés à une Forme Méromorphe. X, 211 pages. 1978.

Vol. 634: H. Kurke et al., Die Approximationseigenschaft lokaler Ringe. IV, 204 Seiten. 1978.

Vol. 635: T. Y. Lam, Serre's Conjecture. XVI, 227 pages. 1978.

Vol. 636: Journées de Statistique des Processus Stochastiques, Grenoble 1977, Proceedings. Edité par Didier Dacunha-Castelle et Bernard Van Cutsem. VII, 202 pages. 1978.

Vol. 637: W. B. Jurkat, Meromorphe Differentialgleichungen. VII, 194 Seiten. 1978.

Vol. 638: P. Shanahan, The Atiyah-Singer Index Theorem, An Introduction. V, 224 pages. 1978.

Vol. 639: N. Adasch et al., Topological Vector Spaces. V, 125 pages. 1978.

continuation on page 203

Lecture Notes in Mathematics

Edited by A. Dold and B. Eckmann

796

Corneliu Constantinescu

Duality in Measure Theory

Springer-Verlag
Berlin Heidelberg New York 1980

Author
Corneliu Constantinescu
Mathematisches Seminar
ETH-Zentrum
8092 Zürich
Switzerland

AMS Subject Classifications (1980):
Primary: 28A33, 28B05, 46E27, 46G10
Secondary: 28A10, 28A25, 28A35, 28C05, 46A20, 46A32, 46A40

ISBN 3-540-09989-1 Springer-Verlag Berlin Heidelberg New York
ISBN 0-387-09989-1 Springer-Verlag New York Heidelberg Berlin

Printing and binding: Beltz Offsetdruck, Hemsbach/Bergstr.
2141/3140-543210

Math Sep.

Table of Contents

Page

Introduction 1

§ 1. Preliminaries

1. Vector lattices 5
2. Measures 6
3. Integration 7
4. Concassage 10
5. Some notations 12
6. Hyperstonian spaces 16

§ 2. Representations

1. Bounded representations of (X,M) 24
2. Representations of measures 26
3. Representations of (X,M) 31
4. Supplementary results concerning the representations 38

§ 3. Duals of spaces of measures

1. Structures on M^{ρ} 45
2. Spaces associated to a measure 53
3. The spaces M, M_b, and M_c 59
4. Structures on M and M_c^{π} 64
5. Spaces of operators 82
6. The spaces M_b and M_b^{π} 89
7. Tensor products of spaces of measures 91
8. The strong D.-P.-property 105
9. The strong approximation property 108

§. 4. Vector measures

1. Preliminaries 114

2. The integral with respect to a vector measure 117

3. Operators on subspaces of M_{c}^{π} 150

4. Vector measures on Hausdorff spaces 161

5. Topologies on the spaces of vector measures 177

Bibliography 191

Index 194

Notations 196

Introduction

The duality in measure theory may be seen as a method to introduce a kind of generalized functions needed in integration theory (with respect to real or vector valued measures).

Let X be a set, let $\underline{\underline{R}}$ be a δ-ring* of subsets of X, let M be the vector lattice of real measures on $\underline{\underline{R}}$, and let M^{π} be the dual of M, i.e. the vector lattice of order continuous linear forms on M. In order to see why the elements of M^{π} are considered as a kind of generalized functions on X, let us denote by F the vector lattice of real functions on X which are universally integrable with respect to any measure of M, i.e.

$$F := \bigcap_{\mu \in M} L^1(\mu) .$$

For any $f \in F$ let $\tilde{f}$ be the map

$$M \to \mathbb{R}, \quad \mu \mapsto \int f d\mu .$$

It is easy to see that $\tilde{f} \in M^{\pi}$ and that

$$F \to M^{\pi}, \; f \mapsto \tilde{f}$$

is a linear map such that $\widetilde{\bigvee_{\iota \in I} f_{\iota}} = \bigvee_{\iota \in I} \tilde{f}_{\iota}$ for any countable family $(f_{\iota})_{\iota \in I}$ in F for which $\bigvee_{\iota \in I} f_{\iota}$ exists. In the most interesting cases the above map is injective but not surjective. The elements of M^{π} which are not of the form $\tilde{f}$ with $f \in F$ may be considered as generalized functions on X. In fact M^{π} does not contain all generalized functions on X furnished by the duality theory but only the small ones, more exactly the universally integrable generalized functions.

In order to obtain all generalized functions, we proceed as follows. Let Φ be the set of fundamental solid subspaces of M. Let $N, N' \in \Phi$ with $N \subset N'$. Then for any $\xi \in N'^{\pi}$ the restriction $\xi|N$ of ξ to N belongs to N^{π} and the map

$$N'^{\pi} \to N^{\pi}, \; \xi \mapsto \xi|N$$

is an imbedding of vector lattices; we identify N'^{π} with a solid subspace of N^{π} via this imbedding. Since $N_1 \cap N_2 \in \Phi$ for any $N_1, N_2 \in \Phi$, the family $(N^{\pi})_{N \in \Phi}$ is upper directed, more exactly an

---

*For the definitions of the used terms see the Index.

inductive system of vector lattices. We set

$$M^{\rho} := \bigcup_{N\in\Phi} N^{\pi} .$$

M^{ρ} is the inductive limit of the inductive system $(N^{\pi})_{N\in\Phi}$, and so it is a complete vector lattice too. M^{ρ} is the set of all generalized functions. In order to see its relation to the set of natural functions, we proceed as above. We denote by G the set of $\underline{\underline{R}}$ - measurable real functions on X, and for any $f\in G$ we denote by $N(f)$ the set of $\mu\in M$ with $f\in L^1(\mu)$. Then $N(f)\in\Phi$. For any $f\in G$ let $\tilde{f}$ be the map

$$N(f) \to \mathbb{R},\ \mu \longmapsto \int f d\mu .$$

It is easy to see that $\tilde{f}\in N(f)^{\pi}$ and therefore $\tilde{f}\in M^{\rho}$. The map

$$G \to M^{\rho} ,\ f \longmapsto \tilde{f}$$

is linear, and we have $\widetilde{\bigvee_{\iota\in I} f_{\iota}} = \bigvee_{\iota\in I} \tilde{f}_{\iota}$ for any countable family $(f_{\iota})_{\iota\in I}$ in G for which $\bigvee_{\iota\in I} f_{\iota}$ exists. Exactly as we remarked above, in the most interesting cases the above map is injective but not surjective. In this way M^{ρ} appears as an extension of the set G of natural measurable functions, but the order properties are improved, since the σ-completeness and the order σ-continuity are replaced by completeness and order continuity. The product of two functions of G belongs to G. This multiplication law on G may be extended in a natural way to M^{ρ} which becomes in this way an associative commutative unital algebra (Theorem 3.1.7).

The space of universally locally integrable generalized functions (which will be denoted by M_c^{π}) is a solid subspace and a unital subalgebra of M^{ρ}, but it possesses in addition a natural topology with respect to which it becomes a complete locally convex lattice (of type M) (Theorem 3.4.8). A similar topology may be defined on M, and M endowed with it and with the natural action $M_c^{\pi} \times M \to M$ becomes a complete locally convex lattice (of type L) (Theorem 3.4.3) and a topological M_c^{π} - module (Theorem 3.4.15). M_c^{π} may be identified with the set of linear maps $\psi: M \to M$ for which $\psi\mu$ is μ-absolutely continuous for any $\mu\in M$ or with the space of linear maps $M \to M$ which are continuous with respect to any canonical seminorm of M (Theorem 3.5.3 and Theorem 3.5.4).

For any continuous real functions φ or $\mathbb{R}$, there exists a map $\hat{\varphi}: M^{\rho} \to M^{\rho}$ which extends in a natural way the map $f \to \varphi\circ f$ defined on

G ; its restriction $M_c^\rho \rightarrow M_c^\rho$ is continuous. This allows us e.g. to define the spaces $L^p(\mu)$ for generalized functions. These results hold for real functions of several variables too (Theorem 3.1.5, 3.4.11, and 3.4.14). The multiplication law in M^ρ quoted above is nothing else but a special case of these considerations. Since the map $\varphi \mapsto \hat{\varphi}$ is continuous (Theorem 3.4.12), we may use the approximation theory for generalized functions exactly like in the classical case.

Let $\mu,\nu \in M$ such that ν is μ-absolutely continuous. The Radon-Nikodym theorem states that if μ is σ-finite (or more generally localizable) then there exists $f \in L^1_{loc}(\mu)$ such that $\nu = f \cdot \mu$. This theorem does not hold any more if we drop the supplementary hypothesis about μ, but it holds even without it if we accept the generalized functions (elements of M^ρ) in $L^1_{loc}(\mu)$ (Theorem 3.2.2). The importance of the Radon-Nikodym theorem lies in the fact that the set $L^1_{loc}(\mu)$, from which the "densities" f are taken, possesses many algebraic and order properties. As it was seen above these properties are even improved by replacing $L^1_{loc}(\mu)$ with M^ρ .

A similar situation occurs in the theory of the space $L^\infty(\mu)$. Namely $L^\infty(\mu)$ is order complete and the dual of $L^1(\mu)$ only if μ possesses a supplementary property (e.g. μ is σ-finite). Again the results hold generally if generalized functions are accepted as members of $L^\infty(\mu)$ (Theorem 3.6.3). Moreover $\{\xi \cdot \mu \mid \xi \in L^\infty(\mu)\}$ is the solid subspace of M generated by μ .

As another application of the duality theory, we construct in 3.7. the topological tensor products of the spaces of measures.

The last section is dedicated to the integration of the generalized function with respect to the vector measures. Since these generalized functions fill the gaps left by the natural functions, they render compact or complete many sets appearing in the integration theory and open by that a new way for simple proofs for some properties concerning the vector measures and the spaces of vector measures. Some applications are given to the theory of operators mapping locally convex spaces of continuous real functions into arbitrary locally convex spaces (generalizations of Riesz's representation theorem).

A word has to be said about the use of δ-rings. Since any σ-ring is a δ-ring, a construction using δ-rings is more general. Any <u>real</u> measure on a σ-ring being bounded, a theory for σ-rings is of little interest (even the Lebesgue measure is not bounded). Let now $\underline{\underline{S}}$ be a ring of sets, let $\underline{\underline{R}}$ be the δ-ring generated by it, and let $M(\underline{\underline{S}}), M(\underline{\underline{R}})$ be the vector lattices of real measures on $\underline{\underline{S}}$ and $\underline{\underline{R}}$ respectively. The map $M(\underline{\underline{R}}) \longrightarrow M(\underline{\underline{S}})$ which assigns to any measure on $\underline{\underline{R}}$ its restriction to $\underline{\underline{S}}$ being an isomorphism of vector lattices, a duality theory for δ-rings is in fact a duality theory for rings of sets. The theory is developed in a frame which allows to apply it to the abstract measures as well as to the Radon measures on Hausdorff spaces (see section 5 of the Preliminaries).

There is a long series of papers in which the duals of the spaces of measures appear ([1], [3], [5], [10], [11], [12], [13], [16], [17], [18], [19], [23], [24], [25], [28], [29], [33]). The present work is logically independent of these papers (i.e. it uses no results from them) and therefore it may be read without any previous knowledge of the duality theory. Some of its results were announced in [6] . A first version of it appeared in a preprint at the Technische Universität Hannover, Institut für Mathematik, Nr. 67, 1977 ([7]) .

The author likes to thank Mrs Rose-Marie Grossmann who expertly carried out the task of typing the manuscript.

§ 1 PRELIMINARIES

In this section we want to specify some terms and notations used in the present paper.

1. Vector lattices

Let E be a vector lattice. We denote by $\overset{E}{\vee}, \overset{E}{\wedge}$ or simple by $\vee, \wedge$ the supremum and the infimum in E respectively and set

$$E_+ := \{x \in E \mid x \geq 0\} ,$$

and

$$x_+ := x \vee 0, \quad x_- := (-x) \vee 0, \quad |x| := x \vee (-x)$$

for any $x \in E$. A subset A of E is called solid if

$$x \in E,\ y \in A,\ |x| \leq |y| \Longrightarrow x \in A$$

A band of E is a solid subspace F of E such that the supremum in E of any subset of F belongs to F if it exists. If any upper bounded nonempty subset of E possesses a supremum in E, we call E an order complete vector lattice.

A linear form x' on E is called positive if it is positive on E_+. A positive linear form x' on E is called order continuous if for any lower directed family $(x_\iota)_{\iota \in I}$ in E with $\bigwedge_{\iota \in I} x_\iota = 0$ we have $\inf_{\iota \in I} x'(x_\iota) = 0$. We denote by $E^+ (E^\pi)$ the subspace of the algebraic dual of E generated by the positive (order continuous positive) linear forms on E. E^+ is an order complete vector lattice and E^π is a band of E^+ and therefore an order complete vector lattice with respect to the induced structure. For any $x \in E$ the map

$$\tilde{x}: E^\pi \to \mathbb{R}, \quad x' \longmapsto \langle x, x' \rangle$$

belongs to $E^{\pi\pi}$. The map

$$E \to E^{\pi\pi} , \quad x \longmapsto \tilde{x}$$

is called the evaluation map.

Assume now E to be an order complete vector lattice. A solid subspace of E is called fundamental if E is the band of E generated by it. By the Riesz theorem a solid subspace F of E is fundamental if any $x \in E$ vanishes if $|x| \wedge |y| = 0$ for any $y \in F$, or if $x = \bigvee_{\substack{y \in F \\ y \leq x}} y$ for any $x \in E_+$. Let Φ be the set of fundamental solid subspaces of E ordered by the converse inclusion relation. It is easy to see that F, $G \in \Phi$ implies $F \cap G \in \Phi$, and therefore Φ is upper directed. Let $F, G \in \Phi$ with $F \subset G$. Then for any $x' \in G^{\pi}$ the restriction $x'|F$ of x' to F belongs to F^{π}. The map

$$G^{\pi} \to F^{\pi}, \qquad x' \mapsto x'|F$$

is an imbedding of vector lattices. We identify G^{π} with its image in F^{π}; by this identification G^{π} becomes a solid subspace of F^{π}. We set $E^{\rho} := \bigcup_{F \in \Phi} F^{\pi}$. E^{ρ} is in fact the inductive limit of $(F^{\pi})_{F \in \Phi}$ and is an order complete vector lattice (see [22]).

A <u>lattice seminorm</u> on E is a seminorm p on E such that

$$|x| \leq |y| \Longrightarrow p(x) \leq p(y)$$

for any $x, y \in E$. An <u>L-seminorm</u> (<u>M-seminorm</u>) is a seminorm on E such that

$$p(|x| + |y|) = p(x) + p(y) \qquad (p(|x| \vee |y|) = \sup(p(x), p(y)))$$

for any $x, y \in E$. Any L-seminorm (M-seminorm) is a lattice seminorm. A <u>locally convex (vector) lattice</u> is a vector lattice endowed with a topology generated by a family of lattice seminorms. A lattice seminorm which is a norm will be called <u>lattice norm</u>; similarly an L-seminorm (M-seminorm) which is a norm will be called an <u>L-norm</u> (<u>M-norm</u>). <u>A normed vector lattice</u> is a vector lattice endowed with a lattice norm; if it is (topologically) complete, it is called a <u>Banach lattice</u>.

2. Measures

A <u>δ-ring</u> is a set $\underline{\underline{R}}$ such that:

a) $\emptyset \in \underline{\underline{R}}$;

b) $\forall A, B \in \underline{\underline{R}} \Rightarrow A \cup B, A \setminus B \in \underline{\underline{R}}$;

c) $\bigcap_{n \in \mathbb{N}} A_n \in \underline{\underline{R}}$ for any sequence $(A_n)_{n \in \mathbb{N}}$ in $\underline{\underline{R}}$.

If only a) and b) are fulfilled, then $\underline{\underline{R}}$ will be called a <u>ring of sets.</u> A <u>real measure</u> or simply a <u>measure</u> on a δ-ring $\underline{\underline{R}}$ is a real function μ on $\underline{\underline{R}}$ such that

$$\mu(\bigcup_{n \in \mathbb{N}} A_n) = \sum_{n \in \mathbb{N}} \mu(A_n)$$

for any disjoint sequence $(A_n)_{n \in \mathbb{N}}$ in $\underline{\underline{R}}$ whose union belongs to $\underline{\underline{R}}$. For any δ-ring $\underline{\underline{R}}$ we denote by $\mathcal{M}(\underline{\underline{R}})$ the set of measures on $\underline{\underline{R}}$. It is a subspace of $\mathbb{R}^{\underline{\underline{R}}}$ and therefore an ordered vector space with respect to the induced structure; in fact it is even an order complete vector lattice.

Let, μ,ν be two measures on a δ-ring $\underline{\underline{R}}$. We say that ν is <u>absolutely continuous with respect to μ</u> and denote it by $\nu \ll \mu$ if

$$\forall A \in \underline{\underline{R}}, \quad |\mu|(A) = 0 \Rightarrow |\nu|(A) = 0 .$$

The set of measures on $\underline{\underline{R}}$ which are absolutely continuous with respect to μ is the band of $M(\underline{\underline{R}})$ generated by μ. More generally we say that ν is absolutely continuous with respect to a subset N of $M(\underline{\underline{R}})$ and denote it by $\nu \ll N$ if ν belongs to the band of $M(\underline{\underline{R}})$ generated by N.

A measure μ on a δ-ring $\underline{\underline{R}}$ is called <u>bounded</u> if

$$\|\mu\| := \sup_{A \in \underline{\underline{R}}} |\mu|(A) < \infty .$$

3. Integration

<u>Throughout this paper we shall denote by X a set and by $\underline{\underline{R}}$ a δ-ring of subsets of X.</u>

Let $\mu \in M(\underline{\underline{R}})$. A subset A of X is called a <u>μ-null set</u> if for any $B \in \underline{\underline{R}}$ there exists $C \in \underline{\underline{R}}$ such that $A \cap B \subset C$ and $|\mu|(C) = 0$. A property concerning the points of X is said to hold <u>μ-almost everywhere</u> (in symbols <u>μ-a.e.</u>) if the set of points where it does not hold is a μ-null set.

For any subset A of X we denote by 1_A^X, or simply 1_A, the characteristic function of A (i.e. the real function on X equal to 1 on A and equal to 0 on $X \setminus A$). A real function f on X is called a <u>step function on X with respect to $\underline{\underline{R}}$</u> if there exists a finite family $(A_\iota)_{\iota \in I}$ in $\underline{\underline{R}}$, and a family $(\alpha_\iota)_{\iota \in I}$ in $\mathbb{R}$ such that

$$f = \sum_{\iota \in I} \alpha_\iota 1_{A_\iota} .$$

We denote by F the set of step functions on X with respect to $\underline{\underline{R}}$, and by F^* the set of functions f on X with values in $\mathbb{R} \cup \{\infty\}$, for which there exists an increasing sequence $(f_n)_{n \in \mathbb{N}}$ in F such that

$$f(x) = \sup_{n \in \mathbb{N}} f_n(x)$$

for any $x \in X$.

Let $\mu \in M(\underline{\underline{R}})_+$. We integrate the functions of F in the usual way and set

$$\int^* f d\mu := \sup\{\int g d\mu \mid g \in F , g \leq f\}$$

for any $f \in F^*$. For any $f \in \bar{\mathbb{R}}^X$ we set

$$\int^* f d\mu := \inf\{\int^* g d\mu \mid g \in F^* , g \geq f \ \mu\text{-a.e.}\} ,$$

$$\int_* f d\mu := -\int^* (-f) d\mu .$$

We say that f is <u>μ-integrable</u> if

$$\int^* f d\mu = \int_* f d\mu \in \mathbb{R} .$$

In this case we set

$$\int f d\mu := \int^* f d\mu .$$

We denote by $L^1(\mu)$ the set of μ-integrable functions.

Let now μ be an arbitrary measure on $\underline{\underline{R}}$. Then

$$L^1(|\mu|) = L^1(\mu_+) \cap L^1(\mu_-) .$$

We set

$$L^1(\mu) := L^1(|\mu|) ,$$

and

$$\int f d\mu := \int f d\mu_+ - \int f d\mu_-$$

for any $f \in L^1(\mu)$. For any $f \in L^1(\mu)$, and for any strictly positive real number α we have $1_{\{f>\alpha\}} \in L^1(\mu)$. For any $f \in \bar{\mathbb{R}}^X$, and for any $A \subset X$ such that $f1_A \in L^1(\mu)$ we set

$$\int_A f d\mu := \int f 1_A d\mu .$$

Let us order $\underline{\underline{R}}$ by the inclusion relation, and let $\underline{\underline{F}}$ be the section filter of $\underline{\underline{R}}$. Then

$$\int f d\mu = \lim_{A,\underline{\underline{F}}} \int_A f d\mu$$

for any $f \in L^1(\mu)$. The following assertions coincide :

a) μ is bounded ;

b) $1_X \in L^1(\mu)$;

c) $\forall A \subset X , (\forall B \in \underline{\underline{R}} \Rightarrow A \cap B \in \underline{\underline{R}}) \Rightarrow 1_A \in L^1(\mu)$.

If these equivalent conditions are fulfilled then

$$\|\mu\| = \int 1_X d|\mu| .$$

We set

$$L^1_{loc}(\mu) := \{f \in \overline{\mathbb{R}}^X \mid \forall A \in \underline{\underline{R}} \Rightarrow f1_A \in L^1(\mu)\} .$$

Any $f \in L^1_{loc}(\mu)$ is finite, μ-a.e. For any upper bounded sequence $(f_n)_{n\in\mathbb{N}}$ in $L^1(\mu)$ $(L^1_{loc}(\mu))$ we have $\bigvee_{n\in\mathbb{N}} f_n \in L^1(\mu)$ $(L^1_{loc}(\mu))$. $L^1(\mu) \cap \mathbb{R}^X$ and $L^1_{loc}(\mu) \cap \mathbb{R}^X$ are subvector lattices of $\mathbb{R}^X$. For any $f \in L^1_{loc}(\mu)$ we denote by $f\cdot\mu$ the map

$$\underline{\underline{R}} \longrightarrow \mathbb{R} , \; A \longmapsto \int_A f d\mu \quad .$$

Then $f\cdot\mu \in M(\underline{\underline{R}})$ and $f\cdot\mu \ll \mu$. We have $|f\cdot\mu| = |f|\cdot|\mu|$ and

$$\mu = \bigvee_{A\in\underline{\underline{R}}} 1_A\cdot\mu$$

if μ is positive. Let $f \in \overline{\mathbb{R}}^X$ and $g \in L^1_{loc}(\mu)$. Then

$$f \in L^1(g\cdot\mu) \Longleftrightarrow fg \in L^1(\mu) \Rightarrow \int (fg) d\mu = \int f d(g\cdot\mu) ,$$

$$f \in L^1_{loc}(g\cdot\mu) \Longleftrightarrow fg \in L^1_{loc}(\mu) \Longrightarrow (fg)\cdot\mu = f\cdot(g\cdot\mu) \text{ *)} .$$

Let $f \in \overline{\mathbb{R}}^X_+$ and let $(\mu_\iota)_{\iota\in I}$ be an upper bounded upper directed family of measures on $\underline{\underline{R}}$ such that $f \in \bigcap_{\iota\in I} L^1(\mu_\iota)$ and

$$\sup_{\iota\in I} \int f d\mu_\iota < \infty \quad .$$

Then $f \in L^1(\bigvee_{\iota\in I} \mu_\iota)$ and

$$\int f d(\bigvee_{\iota\in I} \mu_\iota) = \sup_{\iota\in I} \int f d\mu_\iota \quad .$$

A subset A of X will be called <u>measurable</u> if $A \cap B \in \underline{\underline{R}}$ for any $B \in \underline{\underline{R}}$. A function $f: X \to \overline{\mathbb{R}}$ will be called <u>measurable</u> if $\{f<\alpha\}$ is measurable for any $\alpha \in \mathbb{R}$. A set $\underline{\underline{A}}$ of subsets of X is called <u>locally countable</u> if for any $B \in \underline{\underline{R}}$ the set $\{A \in \underline{\underline{A}} \mid A \cap B \neq \emptyset\}$ is countable. Let $\underline{\underline{A}}$ be a locally countable set of pairwise disjoint measurable subsets of X, and let $(f_A)_{A\in\underline{\underline{A}}}$ be a family in $\overline{\mathbb{R}}^X$. Then the function on X equal to f_A on A for any $A \in \underline{\underline{A}}$, and equal to 0 on $X \setminus \bigcup_{A\in\underline{\underline{A}}} A$ is measurable.

---

*) For any relations P,Q,R we denote by $P \Longleftrightarrow Q \Rightarrow R$, the relation $(P \Longleftrightarrow Q)$ and $(Q \Longrightarrow R)$.

4. Concassage

Let $\mu\in M(\underline{\underline{R}})$. A concassage of μ or a μ-concassage is a set $\underline{\underline{C}}$ of pairwise disjoint sets of $\underline{\underline{R}}$ such that for any $A\in\underline{\underline{R}}$ the set $A\setminus\bigcup_{C\in\underline{\underline{C}}(A)} C$ is a μ-null set, where

$$\underline{\underline{C}}(A) :=\{C\in\underline{\underline{C}} \mid |\mu|(A\cap C)>0\} .$$

Let $N\subset M(\underline{\underline{R}})$. An N-concassage is a set $\underline{\underline{C}}$ which is a μ-concassage for any $\mu\in N$. If N' denotes the band generated by N then any N-concassage is an N'-concassage.

Proposition 1.4.1 *Any bounded measure on* $\underline{\underline{R}}$ *possesses a countable concassage.*

Let μ be a bounded measure on $\underline{\underline{R}}$ and let $(A_n)_{n\in\mathbb{N}}$ be a sequence in $\underline{\underline{R}}$ such that

$$\sup_{n\in\mathbb{N}} |\mu|(A_n) = \|\mu\| .$$

Then $\{A_n\setminus\bigcup_{m=1}^{n-1} A_m \mid n\in\mathbb{N}\}$ is a countable concassage of μ . □

Proposition 1.4.2 *Let* $(N_\iota)_{\iota\in I}$ *be a family of subsets of* $M(\underline{\underline{R}})$ *, and for any* $\iota\in I$ *let* $\underline{\underline{C}}_\iota$ *be a locally countable* N_ι*-concassage such that for any* $A\in\underline{\underline{R}}$ *the set*

$$\{\iota\in I \mid A\cap(\bigcup_{C\in\underline{\underline{C}}_\iota} C)\neq\emptyset\}$$

is countable. Then there exists a locally countable $\bigcup_{\iota\in I} N_\iota$*-concassage.*

Let $\leq$ be a well order on I . We set

$$C_\iota := \bigcup_{C\in\underline{\underline{C}}_\iota} C$$

for any $\iota\in I$ and

$$\underline{\underline{C}} := \bigcup_{\iota\in I}\{C\setminus\bigcup_{\substack{\iota'\in I\\ \iota'<\iota}} C_{\iota'} \mid C\in\underline{\underline{C}}_\iota\} .$$

It is obvious that $\underline{\underline{C}}$ is a locally countable set if pairwise disjoint sets of $\underline{\underline{R}}$. We want to show that $\underline{\underline{C}}$ is a $\bigcup_{\iota\in I} N_\iota$-concassage.

Let $\iota \in I$, let $\mu \in N_\iota$, and let $A \in \underline{\underline{R}}$. We set

$$\underline{\underline{C}}(A) := \{C \in \underline{\underline{C}} \mid |\mu|(A \cap C) > 0\}, \quad \underline{\underline{C}}_\iota(A) := \{C \in \underline{\underline{C}}_\iota \mid |\mu|(A \cap C) > 0\}.$$

We have to show that $A \setminus \bigcup_{C \in \underline{\underline{C}}(A)} C$ is a μ-null set. Since $\underline{\underline{C}}_\iota$ is a μ-concassage, $A \setminus \bigcup_{C \in \underline{\underline{C}}_\iota(A)} C$ is a μ-null set. For any $C \in \underline{\underline{C}} \setminus \underline{\underline{C}}(A)$ the set $A \cap C$ is a μ-null set and

$$\{C \in \underline{\underline{C}} \setminus \underline{\underline{C}}(A) \mid A \cap C \neq \emptyset\}$$

is countable. Hence $\bigcup_{C \in \underline{\underline{C}} \setminus \underline{\underline{C}}(A)} (A \cap C)$ is a μ-null set. From

$$A \setminus \bigcup_{C \in \underline{\underline{C}}(A)} C \subset (A \setminus \bigcup_{C \in \underline{\underline{C}}_\iota(A)} C) \cup (\bigcup_{C \in \underline{\underline{C}} \setminus \underline{\underline{C}}(A)} (A \cap C))$$

it follows that $A \setminus \bigcup_{C \in \underline{\underline{C}}(A)} C$ is a μ-null set. □

<u>*Proposition 1.4.3*</u> *Let N, N' be countable subsets of $M(\underline{\underline{R}})$ such that $|\mu| \wedge |\mu'| = 0$ for any $(\mu, \mu') \in N \times N'$ and such that N' possesses a locally countable concassage. Then there exists a measurable subset A of X such that A is a μ-null set for any $\mu \in N$ and $X \setminus A$ is a μ'-null set for any $\mu' \in N'$.*

Let $\underline{\underline{C}}$ be a locally countable N'-concassage. For any $C \in \underline{\underline{C}}$ there exists $A_C \in \underline{\underline{R}}$ such that $A_C \subset C, A_C$ is a μ-null set for any $\mu \in N$, and $C \setminus A_C$ is a μ'-null set for any $\mu' \in N'$. We set

$$A := \bigcup_{C \in \underline{\underline{C}}} A_C .$$

Then A possesses the required properties. □

<u>*Proposition 1.4.4*</u> *Let $\mu, \nu \in M(\underline{\underline{R}})$ with $\nu \ll \mu$. If ν possesses a locally countable concassage, then there exists a measurable real function f on X such that $f \in \mathcal{L}^1_{loc}(\mu)$ and $\nu = f \cdot \mu$.*

Let $\mu', \mu'' \in M(\underline{\underline{R}})$ such that

$$\mu = \mu' + \mu'' , \quad \mu' \ll \nu, \quad |\mu''| \wedge |\nu| = 0 .$$

Then μ' possesses a locally countable concassage. Since $\nu \ll \mu'$ we deduce the existence of a measurable real function g on X such that $\nu = g \cdot \mu'$. We have $|\mu'| \wedge |\mu''| = 0$, and therefore by Proposition 1.4.3 there exists a measurable subset A of X such that A is a μ''-null set, and $X \setminus A$ is a μ'-null set. The function $f := g 1_A$ possesses the required properties. □

5. Some notations

Throughout this paper we shall denote by X *a set, by* $\underline{\underline{R}}$ *a* δ-*ring of subsets of* X, *and by* M *a band of* $M(\underline{\underline{R}})$. In applications one may take M equal to $M(\underline{\underline{R}})$ obtaining in this way the duality theory for the abstract measures. If X is a Hausdorff space and $\underline{\underline{R}}$ denotes the set of relatively compact Borel sets of X, we may take as M the set of measures on $\underline{\underline{R}}$ which are interior regular with respect to the compact sets. Then M is a band of $M(\underline{\underline{R}})$ and isomorphic as vector lattice to the set of Radon measures on X. Hence in this case we obtain the duality theory for the Radon measures on Hausdorff spaces. Other situations may be considered too (e.g. the spaces of atomical, diffuse, or hyperdiffuse measures). We work therefore with a band of $M(\underline{\underline{R}})$ rather than with $M(\underline{\underline{R}})$ itself in order to obtain a more general theory, which in particular unifies the abstract and the topological case.

We denote by M_b the set of bounded measures of M. M_b is a fundamental solid subspace of M and therefore an order complete vector lattice with respect to the induced structure. The map

$$M_b \to \mathbb{R}_+ , \quad \mu \longmapsto \|\mu\|$$

is an L-norm and M_b endowed with it is an order complete L-space. For any $\mu \in M$, and for any $f \in L^1_{loc}(\mu)$ we have $f \in L^1(\mu)$ iff $f \cdot \mu \in M_b$.

We denote by M_c the set of $\mu \in M$ for which there exists $A \in \underline{\underline{R}}$ such that $X \setminus A$ is a μ-null set. M_c is a fundamental solid subspace of M and of M_b.

For any $f \in \overline{\mathbb{R}}^X$ we set

$$M(f) := \{\mu \in M \mid f \in L^1(\mu)\}$$

and denote by $\dot{f}$ the map

$$M(f) \to \mathbb{R} , \quad \mu \mapsto \int f d\mu .$$

$M(f)$ is a solid subspace of M. We denote by L_∞ the set of $f \in \mathbb{R}^X$ for which $M(f)$ is fundamental. We set

$$L := \mathbb{R}^X \cap (\bigcap_{\mu \in M} L^1(\mu)) \quad , \quad L_b := \mathbb{R}^X \cap (\bigcap_{\mu \in M_b} L^1(\mu)) \quad ,$$

$$L_c := \mathbb{R}^X \cap (\bigcap_{\mu \in M_c} L^1(\mu)) \ .$$

We have

$$L \subset L_b \subset L_c \subset L_\infty \ .$$

<u>*Proposition 1.5.1*</u> *Let* $f \in \bar{\mathbb{R}}^X$ *such that* $M(f)$ *is a fundamental solid subspace of* M. *Then* $\{|f|=\infty\}$ *is a* μ*-null set for any* $\mu \in M$, *and any real function on* X *equal to* f *on* $\{|f|<\infty\}$ *belongs to* L_∞.

For any $\nu \in M(f)$ the set $\{|f|=\infty\}$ is a ν-null set. Since $M(f)$ is a fundamental solid subspace of M, we have

$$|\mu| = \bigvee_{\substack{\nu \in M(f) \\ \nu \leq |\mu|}} \nu \ ,$$

and therefore $\{|f|=\infty\}$ is a μ-null set.

Let g be a real function on X equal to f on $\{|f|<\infty\}$. By the above result $f=g$, μ-a.e. for any $\mu \in M$, and therefore $M(g) = M(f)$. Hence $g \in L_\infty$. □

<u>*Proposition 1.5.2*</u> L_∞ *is a unital subalgebra of* $\mathbb{R}^X$ *and a subvector lattice of* $\mathbb{R}^X$ *such that the function*

$$X \longmapsto \mathbb{R} \ , \quad x \longmapsto \sup_{n \in \mathbb{N}} f_n(x)$$

belongs to L_∞ *for any upper bounded sequence* $(f_n)_{n \in \mathbb{N}}$ *in* L_∞. *Any measurable real function on* X *belongs to* L_∞.

Let $f,g \in L_\infty$, and let $\alpha,\beta \in \mathbb{R}$. We set

$$N := \{|\mu| \wedge |\nu| \mid (\mu,\nu) \in M(f) \times M(g)\} \ .$$

Then N is a fundamental solid subspace of M. Since

$$N \subset M(\alpha f + \beta g) \cap M(f \vee g) \ ,$$

it follows that $\alpha f + \beta g, f \vee g \in L_\infty$. Hence L_∞ is a subvector lattic of $\mathbb{R}^X$. We denote by h the function on X equal to $\frac{1}{|g|}$ on

$\{|g|>1\}$, and equal to 1 elsewhere. Then $h \in \bigcap_{\mu\in M} L_{loc}(\mu)$ and $\{h\cdot\mu \mid \mu\in M(f)\}$ is a fundamental solid subspace of M. From

$$\{h\cdot\mu \mid \mu\in M(f)\} \subset M(fg)$$

we get $fg \in L_\infty$, and therefore L_∞ is a unital subalgebra of $\mathbb{R}^X$.

Let $(f_n)_{n\in\mathbb{N}}$ be an upper bounded sequence in L_∞, and let f be the function

$$X \longrightarrow \mathbb{R}, \quad x \longmapsto \sup_{n\in\mathbb{N}} f_n(x).$$

Since L_∞ is a subvector lattice of $\mathbb{R}^X$, we may assume that there exists $g \in L_\infty$ such that $|f_n| \leq g$ for any $n \in \mathbb{N}$. Then $M(g) \subset M(f)$ and there $f \in L_\infty$.

Let f be a measurable real function on X and let $\mu \in M_b$. The function g on X equal to $\frac{1}{f}$ on $\{|f|>1\}$, and equal to 1 on $\{|f|\leq 1\}$ is measurable and bounded, and therefore belongs to $L^1(\mu)$. The same being true for fg, we have $fg \in L^1(\mu)$. Hence $f \in L^1(g\cdot\mu)$. Since $\mu << g\cdot\mu$, and since μ is arbitrary, we get $f \in L_\infty$. □

Proposition 1.5.3

a) L, L_b *and* L_c *are solid subspaces of* L_∞;
b) L_b *and* L_c *are unital subalgebras of* L_∞;
c) L *is an ideal of* L_c;
d) any bounded function of L_∞ *belongs to* L_b;
e) $L_c = \bigcap_{\mu\in M} L^1_{loc}(\mu)$.

a) is obvious.

b) Let $f,g \in L_c$, and let $\mu \in M_c$. Then $g\cdot\mu \in M_c$, and we deduce $f \in L^1(g\cdot\mu)$ and $fg \in L^1(\mu)$. Hence $fg \in L_c$. A similar proof shows that $f,g \in L_b$ implies $fg \in L_b$. Since $1_X \in L_b$ it follows that L_b and L_c are unital subalgebras of L_∞.

c) Let $(f,g)\ L \times L_c$, and let $\mu \in M$. Then $g \in L^1_{loc}(\mu)$, and therefore $f \in L^1(g\cdot\mu)$ and $fg \in L^1(\mu)$. Hence $fg \in L$, and L is an ideal of L_c.

d) Let f be a bounded function of L_∞. In order to show that $f \in L_b$ we may assume f positive. Let $\mu \in M_{b+}$. Then $\{\nu \in M(f) \mid \nu \leq \mu\}$ is upper directed and μ is its supremum. Moreover since f is bounded

$$\sup_{\substack{\nu\in M(f)\\ \nu\leq\mu}} \int f d\nu \leq \int^* f d\mu < \infty$$

and therefore $f \in L^1(\mu)$.

e) is obvious. □

Proposition 1.5.4 *For any* $f\in L_\infty$ *and for any Borel set* B *of* $\mathbb{R}$ *we have* $1_{f^{-1}(B)} \in L_b$.

For any strictly positive real number α we have $M(1_{\{f>\alpha\}}) \supset M(f)$ and therefore $1_{\{f>\alpha\}} \in L_\infty$. By Proposition 1.5.3 d) $1_{\{f>\alpha\}} \in L_b$. We deduce $1_{\{f<-\alpha\}} \in L_b$. The set of subsets B of $\mathbb{R}$ for which $1_{f^{-1}(B)} \in L_b$ being a σ-algebra it contains any Borel set of $\mathbb{R}$. □

Proposition 1.5.5

a) $\dot{f} \in M(f)^\pi \subset M^\rho$ *for any* $f \in L_\infty$;

b) *the map*

$$L_\infty \longrightarrow M^\rho, \quad f \longmapsto \dot{f}$$

is a linear map such that

$$\widehat{\bigvee_{n\in\mathbb{N}} f_n}^{\,\cdot} = \bigvee_{n\in\mathbb{N}} \dot{f}_n$$

for any upper bounded sequence $(f_n)_{n\in\mathbb{N}}$ *in* L_∞ *and such that*

$$\dot{f} = \bigvee_{A\in\underline{\underline{R}}} \widehat{f 1_A}^{\,\cdot}$$

for any $f \in L_{\infty+}$;

c) *we have*

$$f\in L_c \Leftrightarrow \dot{f}\in M_c^\pi \quad ,$$
$$f\in L_b \Leftrightarrow \dot{f}\in M_b^\pi \quad ,$$
$$f\in L \Leftrightarrow \dot{f}\in M^\pi$$

for any $f \in L_\infty$;

d) *if* $\underline{\underline{S}}$ *is a subset of* $\underline{\underline{R}}$ *such that the union of any sequence in* $\underline{\underline{S}}$ *belongs to* $\underline{\underline{S}}$ *then* $\bigvee_{A\in\underline{\underline{S}}} \dot{1}_A \in M^\pi$.

a) is obvious.

b) It is obvious that the considered map is linear. Let $f,g \in L_\infty$ and let $\mu \in M(f) \cap M(g)$, $\mu \geq 0$. We have $1_{\{f<g\}}, 1_{\{g<f\}} \in L^1_{loc}(\mu)$ and

$$\dot{\widehat{f \vee g}}(\mu) = \int (f \vee g) d\mu = \int f d(1_{\{g \leq f\}} \cdot \mu) + \int g d(1_{\{f<g\}} \cdot \mu) =$$

$$= \dot{f}(1_{\{g \leq f\}} \cdot \mu) + \dot{g}(1_{\{f<g\}} \cdot \mu) \leq (\dot{f} \vee \dot{g})(\mu) \leq \dot{\widehat{f \vee g}}(\mu) .$$

Hence $\dot{\widehat{f \vee g}} = \dot{f} \vee \dot{g}$. Let now $(f_n)_{n \in \mathbb{N}}$ be an upper bounded sequence in L_∞. In order to prove the desired assertion we may assume by the above considerations $(f_n)_{n \in \mathbb{N}}$ increasing. By Proposition 1.5.2 $\bigvee_{n \in \mathbb{N}} f_n$ exists in L_∞ and by Lebesgue convergence theorem

$$\dot{\widehat{\bigvee_{n \in \mathbb{N}} f_n}} = \bigvee_{n \in \mathbb{N}} \dot{f}_n .$$

Let $f \in L_{\infty +}$. Then

$$\dot{f}(\mu) = \int f d\mu = \sup_{A \in \underline{\underline{R}}} \int_A f d\mu = \sup_{A \in \underline{\underline{R}}} \dot{\widehat{f 1_A}}(\mu)$$

for any $\mu \in M(f)_+$ and therefore $\dot{f} = \bigvee_{A \in \underline{\underline{R}}} \dot{\widehat{f 1_A}}$.

c) is obvious.

d) follows immediately from the relation

$$\sup_{A \in \underline{\underline{S}}} |\mu|(A) < \infty$$

for any $\mu \in M$. □

6. Hyperstonian spaces

Let Y be a locally compact hyperstonian space and let $M(Y)$ be the set of normal Radon measures on Y ([9]). $M(Y)$ is isomorphic as vector lattice to a band of the space of measures defined on the set of relatively compact Borel sets of Y and so we may use the notations and results of the preceding section with respect to $M(Y)$. In order to distinguish this special case, which plays an important part in this paper, we shall add (Y) to all symbols defined with the aid of $M(Y)$. In particular $M_b(Y)$ will stay for the set of bounded measures of $M(Y)$ and $M_c(Y)$ for the set of measures of $M(Y)$ with compact support.

We denote by $C_\infty(Y)$ the set of continuous $f \in \overline{\mathbb{R}}^X$ for which

$\{|f| = \infty\}$ is nowhere dense. For any $f,g \in C_\infty(Y)$ there exists a unique function of $C_\infty(Y)$, which will be denoted by $f + g$, such that

$$(f+g)(x) = f(x) + g(x)$$

for any $x \in \{|f|<\infty\} \cap \{|g|<\infty\}$. In a similar way we defined fg and αf, where $\alpha \in \mathbb{R}$. $C_\infty(Y)$ endowed with the corresponding structure is an associative, commutative, unital algebra. The relation

$$f \leqslant g :\Longleftrightarrow (\forall x \in X \Longrightarrow f(x) \leqslant g(x))$$

is an order relation on $C_\infty(Y)$ and $C_\infty(Y)$ endowed with it becomes an order complete vector lattice ([27] 26.2.2). We denote by $C(Y)$ the set of finite functions of $C_\infty(Y)$ (i.e. the set of continuous real functions on Y), by $C_b(Y)$ the set of bounded functions of $C(Y)$, by $C_i(Y)$ the set of functions of $C(Y)$ which are μ-integrable for any $\mu \in M(Y)$ and by $C_c(Y)$ the set of functions of $C(Y)$ with compact support. $C(Y)$ is a subalgebra of $C_\infty(Y)$, $C_b(Y)$ is a subalgebra of $C(Y)$ and $C_i(Y)$, $C_c(Y)$ are ideals of $C(Y)$. For any $\mu \in M(Y)$ and for any $f \in C_\infty(Y)$ we denote by $\mathrm{Supp}_Y\, \mu$ and $\mathrm{Supp}_Y\, f$ or simply by $\mathrm{Supp}\, \mu$ and $\mathrm{Supp}\, f$ the support of μ and f respectively; if Y' is an open subspace of Y then $\mu|Y'$ and $f|Y'$ denote the restriction of μ and f to Y' respectively.

Proposition 1.6.1

a) *The map*

$$C_\infty(Y) \longrightarrow M(Y)^\rho, \quad f \longmapsto \dot{f}$$

is an isomorphism of vector lattices ;

b) $\{\dot{f} | f \in C(Y)\} = M_c(Y)^\pi$, $\{\dot{f} | f \in C_b(Y)\} = M_b(Y)^\pi$,
$\{\dot{f} | f \in C_i(Y)\} = M(Y)^\pi$

and the maps

$$C(Y) \longrightarrow \dot{M}_c(Y), \quad f \longmapsto \dot{f},$$
$$C_b(Y) \longrightarrow \dot{M}_b(Y), \quad f \longmapsto \dot{f},$$
$$C_i(Y) \longrightarrow \dot{M}(Y) \quad f \longmapsto \dot{f}$$

are isomorphisms of vector lattices ;

c) *the evaluation maps* $M(Y) \longrightarrow M(Y)^{\pi\pi}$, $M_b(Y) \longrightarrow M_b(Y)^{\pi\pi}$ *are isomorphisms of vector lattices ;*

d) *the evaluation map* $M_c(Y) \longrightarrow M_c(Y)^{\pi\pi}$ *is injective and its image is a fundamental solid subspace of* $M_c(Y)^{\pi\pi}$.

a) Let $\xi \in M(Y)^{\rho}$ and let N be a fundamental solid subspace of $M(X)$ such that $\xi \in N^{\pi}$. Then for any $\mu \in N$ the map

$$L^{\infty}(\mu) \longrightarrow \mathbb{R} \ , \ g \longrightarrow \xi(g\cdot\mu)$$

belongs to $(L^{\infty}(\mu))^{\pi}$ and therefore there exists $f_{\mu} \in C_{\infty}(Y) \cap L^1(\mu)$ such that

$$\int f_{\mu} g d\mu = \xi(g\cdot\mu)$$

for any $g \in L^{\infty}(\mu)$ ([9] Propositions 2.1 and 2.2 and N. Bourbaki 2ème ed. Intégration Ch V, § 5 Proposition 14). Let $\mu, \nu \in N$ with $|\nu| \leqslant |\mu|$. Then there exists $g \in L^{\infty}(\mu)$ such that $g\cdot\mu = \nu$ and we get for any $h \in L^{\infty}(\nu)$

$$\int f_{\nu} h d\nu = \xi(h\cdot\nu) = \xi((hg)\cdot\mu) = \int f_{\mu}(hg)d\mu = \int f_{\mu} h d\nu \quad .$$

Hence $f_{\mu} = f_{\nu}$ on Supp ν . For arbitrary $\mu, \nu \in N$ we deduce $f_{\mu} = f_{\nu}$ on Supp $\mu \cap$ Supp ν . Hence there exists $f \in C_{\infty}(Y)$ such that

$$\dot{f}(\mu) = \int f d\mu = \xi(\mu)$$

for any $\mu \in N$. We get $\dot{f} = \xi$ and therefore the map

$$C_{\infty}(Y) \longrightarrow M(Y)^{\rho} \ , \quad f \longmapsto \dot{f}$$

is surjective. It is obviously an injection and by Proposition 1.5.5 b) it is an isomorphism of vector lattices.

b) Let $f \in C_{\infty}(Y) \cap L_c(Y)$ and let $x \in X$ such that $|f(x)| = \infty$. Then there exists a compact neighbourhood K of x and a sequence $(\mu_n)_{n \in \mathbb{N}}$ in $M(Y)_+$ with

$$\text{Supp}\ \mu_n \subset K \cap \{n \leqslant |f|\} \ , \quad \mu_n(K)=1$$

for any $n \in \mathbb{N}$. We get

$$\mu := \sum_{n \in \mathbb{N}} \frac{1}{n^2} \mu_n \in M_c(Y)$$

and

$$\infty > |\dot{f}|(\mu) = \int |f| d\mu = \sum_{n \in \mathbb{N}} \frac{1}{n^2} \int |f| d\mu_n \geq \sum_{n \in \mathbb{N}} \frac{1}{n} = \infty .$$

Hence $\mathcal{C}_\infty(Y) \cap \mathcal{L}_c(Y) = \mathcal{C}(Y)$. The relations

$$\mathcal{C}_\infty(Y) \cap \mathcal{L}_b(Y) = \mathcal{C}_b(Y) \ , \ \mathcal{C}_\infty(Y) \cap \mathcal{L}(Y) = \mathcal{C}_i(Y)$$

are obvious. The assertions follow now from a) and Proposition 1.5.5 c).

c) follows immediately from b) .

d) Let Y^* be the Stone-Čech compactification of Y . Then Y^* is a hyperstonian space, $\mathcal{C}(Y)$ may be identified with

$$\{f \in \mathcal{C}_\infty(Y^*) \mid |f| < \infty \text{ on } Y\} \ ,$$

$\mathcal{C}(Y)^\pi$ with

$$\{\mu \in \mathcal{M}(Y^*) \mid \mathcal{C}(Y) \subset \mathcal{L}^1(\mu)\} \ ,$$

and $\mathcal{M}_c(Y)$ with

$$\{\mu \in \mathcal{M}(Y^*) \mid \operatorname{Supp} \mu \subset Y\} \ .$$

The assertion follows immediately from these relations and b). □

<u>Remark</u> The evaluation map $\mathcal{M}_c(Y) \longrightarrow \mathcal{M}_c(Y)^{\pi\pi}$ is not always an isomorphism. Indeed let $\mathcal{N}^*$ be the Stone-Čech compactification of $\mathcal{N}$, let x be a point of $\mathcal{N}^* \setminus \mathcal{N}$, and let Y be $\mathcal{N}^* \setminus \{x\}$. Then $\mathcal{C}(Y) = \mathcal{C}_b(Y)$ and therefore $\mathcal{M}_b(Y)^{\pi\pi} = \mathcal{M}_c(Y)^{\pi\pi}$, while $\mathcal{M}_c(Y) \neq \mathcal{M}_b(Y)$.

<u>*Proposition 1.6.2*</u> *Let* $f \in \mathcal{C}_i(Y)$ *and let* $(U_n)_{n \in \mathbb{N}}$ *be a decreasing sequence of open sets of* Y *such that* $\bar{U}_n \cap \operatorname{Supp} f \neq \emptyset$ *for any* $n \in \mathbb{N}$. *Then* $(\bigcap_{n \in \mathbb{N}} \bar{U}_n) \cap \operatorname{Supp} f \neq \emptyset$. *In particular any continuous real function on* Supp f *is bounded and*

$$\{f=0\} \cap \operatorname{Supp} f = \emptyset \Longrightarrow \inf_{x \in \operatorname{Supp} f} |f(x)| > 0 \text{ and } 1_{\operatorname{Supp} f} \in \mathcal{C}_i(Y) \ .$$

Assume $(\bigcap_{n \in \mathbb{N}} \bar{U}_n) \cap \operatorname{Supp} f = \emptyset$. Then for any compact set K of Y

there exists $n \in \mathbb{N}$ such that $K \cap \bar{U}_n \cap \mathrm{Supp}\, f = \emptyset$. For any $n \in \mathbb{N}$ there exists $\mu_n \in M_c(Y)_+$ with

$$\mathrm{Supp}\, \mu_n \subset \bar{U}_n \cap \mathrm{Supp}\, f \ , \ \int |f| d\mu_n = 1 \ .$$

By the above remark $(\mu_n)_{n \in \mathbb{N}}$ is summable in $M(Y)$ and we get the contradictory relation

$$\infty > \int |f| d(\sum_{n \in \mathbb{N}} \mu_n) = \sum_{n \in \mathbb{N}} \int |f| d\mu_n = \infty \ .$$

If g is a continuous real function on Supp f then $\{|g|>n\}$ is a decreasing sequence of open sets of Y such that $\bigcap_{n \in \mathbb{N}} \{|g|>n\} = \emptyset$. Hence there exists $n \in \mathbb{N}$ such that $|g| \leq n$.

If $\{f = 0\} \cap \mathrm{Supp}\, f = \emptyset$ then

$$\mathrm{Supp}\, f \longrightarrow \mathbb{R} \ , \ x \longmapsto \frac{1}{f(x)}$$

is a continuous real function on Supp f and therefore bounded. Hence $\inf_{x \in \mathrm{Supp}\, f} |f(x)| > 0$ and $1_{\mathrm{Supp}\, f} \in C_i(Y)$. □

Remark The fact that any continuous real function on Supp f is bounded was proved by S. Kaplan ([16]III (7.8)).

Proposition 1.6.3 *If $M(Y) = M_b(Y)$ then $C(Y) = C_b(Y)$.*

If $M(Y) = M_b(Y)$ then $1_Y \in C_i(Y)$ and by Proposition 1.6.2 any continuous real function on Y is bounded. □

Proposition 1.6.4 *Let Y^* be the Stone-Čech compactification of Y, let F be the set $\{f \in C(Y^*) \mid (f|Y) \in C_i(Y)\}$, and let Z be an open set of Y^* containing Y. Then the following assertions are equivalent:*

a) $Z \subset \bigcup_{f \in F} \{f \neq 0\}$;

b) *the map*

$$M(Z) \longrightarrow M(Y) \ , \ \mu \longrightarrow \mu|Y$$

is surjective (and therefore bijective).

a $\Longrightarrow$ b. Let $\nu \in M(Y)$ and let K be a compact set of Z. By a) there exists $f \in F$ such that $K \subset \{f \neq 0\}$. We set

$$\alpha := \inf_{y \in K} |f(y)| > 0 .$$

Then $1_K^{Y^*} \leq \frac{1}{\alpha} |f|$ and therefore $1_{K \cap Y}^{Y} \in L^1(\nu)$. Since K is arbitrary there exists $\mu \in M(Z)$ such that $\mu | Y = \nu$. Hence the map

$$M(Z) \longrightarrow M(Y) , \mu \longmapsto \mu | Y$$

is surjective.

b $\Longrightarrow$ a. Let $f \in C(Y^*)$ with Supp $f \subset Z$ and let $\nu \in M(Y)$. By b) there exists $\mu \in M(Z)$ such that $\mu | Y = \nu$. Since $f \in L^1(\mu)$ we deduce $f|Y \in L^1(\nu)$ and $f \in F$. Hence $Z \subset \bigcup_{f \in F} \{f \neq 0\}$. □

For any $g \in C_i(Y)$ we denote by p_g the map

$$C(Y) \longrightarrow \mathbb{R}_+ , f \longrightarrow \inf \{\alpha \in \mathbb{R}_+ | \ |fg| \leq \alpha |g|\} = \sup_{x \in \text{Supp } g} |f(x)| .$$

Proposition 1.6.5

a) p_g *is an M-seminorm for any* $g \in C_i(Y)$, *and we have*

$$p_g(f f') \leq p_g(f) \, p_g(f')$$

for any $f, f' \in C(Y)$;

b) $C(Y)$ *endowed with the topology generated by the family* $(p_g)_{g \in C_i(X)}$ *of seminorms is a complete locally convex space and a topological ring ;*

c) $C_b(Y)$ *is a dense set of* $C(Y)$ *with respect to the above topology.*

p_g is a seminorm by Proposition 1.6.2 and it is obviously an M-seminorm for any $g \in C_i(Y)$. The other assertions are obvious. □

Proposition 1.6.6 *Let* Y^* *be the Stone-Čech compactification of* Y, *let* φ *be a positive linear form on* $C_\infty(Y)$, *and for any* $f \in C_\infty(Y)$ *let* f^* *be its continuous extention to* Y^*. *Then :*

a) there exists a finite subset M *of* Y^* *and a positive real* *function* h *on* M *such that*

$$\varphi f = \sum_{y\in M} h(y) f^*(y)$$

for any $f \in C_\infty(Y)$;

b) $\varphi \in C_\infty(Y)^\pi$ *iff any point of* M *is an isolated point of* Y^* *(in which case* $M \subset Y$*)* .

a) The restriction of φ to $C_b(Y)$ is a positive linear form and therefore there exists a positive Radon measure λ on Y^* such that

$$\varphi f = \int f^* d\lambda$$

for any $f \in C_b(Y)$.

Let $f \in C_\infty(Y)_+$. Then $f \wedge n 1_Y \in C_b(Y)$ and therefore

$$\varphi f \geqslant \int f^* \wedge n 1_{Y^*} d\lambda$$

for any $n \in \mathbb{N}$. Hence

$$\varphi f \geqslant \int f^* d\lambda \quad .$$

For any continuous map $\psi : \bar{\mathbb{R}} \longrightarrow \bar{\mathbb{R}}_+$ with $\psi(\mathbb{R}) \subset \mathbb{R}_+$ we have $\psi \circ f^* \in L^1(\lambda)$ and therefore f^* is bounded on Supp λ. Since

$$f^2 \geqslant n(f - f \wedge n 1_Y)$$

for any $n \in \mathbb{N}$ we get

$$\varphi(f^2) \geqslant n(\varphi f - \int f^* d\lambda)$$

for any $n \in \mathbb{N}$ with

$$n \geqslant \sup_{y \in \text{Supp}\,\lambda} f^*(y) \quad .$$

Hence

$$\varphi f = \int f^* d\lambda$$

and this relation holds for any $f \in C_\infty(Y)$.

Assume Supp λ is not finite. Then there exists a disjoint sequence $(K_n)_{n\in\mathbb{N}}$ of compact open sets of Y^* such that

$$K_n \cap \operatorname{Supp} \lambda \neq \emptyset$$

for any $n \in \mathbb{N}$. There exists $f \in \mathcal{C}_\infty(Y)_+$ such that

$$f \geqslant \frac{1}{\lambda(K_n \cap \operatorname{Supp} \lambda)}$$

for any $n \in \mathbb{N}$ and we get the contradictory relation

$$\infty > \varphi f = \int f^* d\lambda \geqslant \sum_{n \in \mathbb{N}} \int_{K_n} f d\lambda \geqslant \sum_{n \in \mathbb{N}} 1 = \infty .$$

Hence Supp λ is finite and therefore a) holds.

b) is obvious. □

§ 2 REPRESENTATIONS

1. Bounded representation of (X,M)

Definition 2.1.1 *An ordered triple* (Y,u,v) *is called a bounded representation of* (X,M) *if :*

a) Y *is a compact hyperstonian space ;*

b) u *is a homomorphism of unital algebras* $L_b \to C_b(Y)$ *such that*

$$u(\bigvee_{n\in\mathbb{N}} f_n) = \bigvee_{n\in\mathbb{N}} u(f_n)$$

for any upper bounded sequence $(f_n)_{n\in N}$ *in* L_b *;*

c) v *is an isomorphism of normed vector lattices* $M_b \to M_b(Y)$ *;*

d) *we have*

$$\int f d\mu = \int (uf) d(v\mu) \ , \ v(f\cdot\mu) = (uf)\cdot(v\mu)$$

for any $(f,\mu)\in L_b\times M_b$.

Theorem 2.1.2 *Let* (Y,u,v) , (Y',u',v') *be bounded representations of* (X,M) . *Then there exists a unique homeomorphism* $\varphi: Y \to Y'$ *such that* $v'\mu = \varphi(v\mu)$ *for any* $\mu\in M_c$. *We have* $u f = (u'f)\circ\varphi$ *for any* $f\in L_b$ *and* $v'\mu = \varphi(v\mu)$ *for any* $\mu\in M_b$.

The map

$$v'\circ v^{-1}: M_b(Y) \longrightarrow M_b(Y')$$

is an isomorphism of normed vector lattices. Let

$$w : C_b(Y') \longrightarrow C_b(Y)$$

be its adjoint map (Proposition 1.6.1 b)). It is an isomorphism of normed vector latices too. By [27] Theorem 7.8.4 there exists a homeomorphism $\varphi: Y \to Y'$ such that $wg' = g'\circ\varphi$ for any $g'\in C_b(Y')$. We get

$$\int g' d\varphi(v\mu) = \int g'\circ\varphi \, d(v\mu) = \int (wg') d(v\mu) = \int g' d(v'\mu)$$

for any $(g',\mu)\in C_b(Y')\times M_b$ and therefore $v'\mu = \varphi(v\mu)$ for any $\mu\in M_b$. We have

$$\int (u'f)\circ\varphi \, d(v\mu) = \int (u'f) d\varphi(v\mu) = \int (u'f) d(v'\mu) = \int f d\mu = \int (uf) d(v\mu)$$

for any $(f,\mu)\in L_b\times M_b$ and therefore $u f = (u'f)\circ\varphi$ for any $f\in L_b$.

The uniqueness of φ is obvious. □

<u>*Theorem 2.1.3*</u> *There exists a bounded representation for* (X,M) .

By [26] Theorem V 8.6 there exist a compact hyperstonian space Y and an isomorphism $v : M_b \longrightarrow M_b(Y)$ of normed vector lattices. For any $f \in L_b$ the map

$$M_b(Y) \longrightarrow \mathbb{R} \,,\; \lambda \longrightarrow \int f d(v^{-1}\lambda)$$

belongs to $M_b(Y)^{\pi}$. By Proposition 1.6.1 a), b) there exists a unique $uf \in C_b(Y)$ such that

$$\int f d\mu = \int (uf) d(v\mu)$$

for any $\mu \in M_b$. It is obvious that $u : L_b \longrightarrow C_b(Y)$ is linear.

Let $(f_n)_{n\in\mathbb{N}}$ be an upper bounded sequence in L_b and let $\lambda \in M_b(Y)_+$. We have

$$\int u(\bigvee_{n\in\mathbb{N}} f_n) d\lambda = (\bigvee_{n\in\mathbb{N}} f_n) d(v^{-1}\lambda) =$$

$$= \sup\{ \sum_{n\in M} \int f_n d\mu_n \mid M \text{ finite} \subset \mathbb{N},\ (\mu_n)_{n\in M} \text{ family in } M_{b+},\ \sum_{n\in M} \mu_n = (v^{-1}\lambda)\} =$$

$$= \sup\{ \sum_{n\in M} \int (uf_n) d\lambda_n \mid M \text{ finite} \subset \mathbb{N}, (\lambda_n)_{n\in M} \text{ family in } M_b(Y)_+, \sum_{n\in M} \lambda_n = \lambda\} =$$

$$= \int (\bigvee_{n\in\mathbb{N}} (uf_n)) d\lambda \quad .$$

Since λ is arbitrary we get $u(\bigvee_{n\in\mathbb{N}} f_n) = \bigvee_{n\in\mathbb{N}} (uf_n)$. In particular u is an homomorphism of vector lattices.

We have

$$\int (u\, 1_X) d\lambda = \int 1\, d(v^{-1}\lambda) = \|v^{-1}\lambda\| = \|\lambda\| = \int 1_Y d\lambda$$

for any $\lambda \in M_b(Y)_+$ and therefore $u\, 1_X = 1_Y$.

Let $f,g \in u(L_b)$ and let $f',f'',g',g'' \in L_b$ with

$$uf' = uf'' = f \,,\quad ug' = ug'' = g.$$

Then for any $\mu\in M_b$ we have $f' = f''$ μ-a.e. and $g' = g''$ μ-a.e. and therefore $f'g' = f''g''$ μ-a.e. Hence $u(f'g') = u(f''g'')$. We set $F(f,g) := u(f'g')$. Then F is a map

$$u(L_b) \times u(L_b) \longrightarrow u(L_b)$$

such that

$$F(f+g,h) = F(f,h) + F(g,h) \ , \ F(f,g) = F(g,f) \ ,$$

$$F(f,1) = f$$

$$f \geqslant 0 \ , \ g \geqslant 0 \Longrightarrow F(f,g) \geqslant 0$$

for any $f,g,h \in u(L_b)$. By [32] Theorem V 8.2 we get $F(f,g) = fg$ for any $f,g \in u(L_b)$. This shows that u is a homomorphism of unital algebras.

Let $(f,\mu) \in L_b \times M_b$. We have for any $A \in \underline{\underline{R}}$

$$(f\cdot\mu)(A) = \int f1_A d\mu = \int u(f1_A)d(v\mu) = \int (uf)(u1_A)d(v\mu) =$$

$$= \int (u1_A)d((uf)\cdot(v\mu)) = \int 1_A d(v^{-1}((uf)\cdot(v\mu))) = v^{-1}((uf)\cdot(v\mu))(A)$$

and therefore $f\cdot\mu = v^{-1}((uf)\cdot(v\mu))$, $v(f\cdot\mu) = (uf)\cdot(v\mu)$. □

2. Representations of measures

Definition 2.2.1 *Let* $\mu\in M$, *let* M_μ *be the band of* M *generated by* μ , *and let* (Y,u,v) *be a bounded representation of* (X,M) . *A representation of* μ *associated to* (Y,u,v) *is an ordered triple* (Y_o,u_o,v_o) *such that :*

a) Y_o *is a closed and open set of* $\bigcup_{A\in\underline{\underline{R}}} \mathrm{Supp}\,(u1_A)$;

b) v_o *is an isomorphism of vector lattices* $M_\mu \longrightarrow M(Y_o)$ *such that* $v_o\nu = (v\nu)|Y_o$ *for any* $\nu\in M_\mu \cap M_b$;

c) u *is a map* $L^1_{loc}(\mu) \longrightarrow C_\infty(Y_o) \cap L^1_{loc}(v_o\mu)$ *such that*

$$(u_o f)\cdot(v_o\mu) = v_o(f\cdot\mu)$$

for any $f\in L^1_{loc}(\mu)$.

Proposition 2.2.2 Let $\mu \in M$, let M_μ be the band of M generated by μ, and let (Y,u,v) be a bounded representation of (X,M). Then there exists a unique representation (Y_o,u_o,v_o) of μ associated to (Y,u,v) and we have :

a) $Y_o = \bigcup_{A \in \underline{\underline{R}}} \text{Supp } v(1_A \cdot \mu)$;

b) $Y_o = \text{Supp } (v_o \cdot \mu)$;

c) $u_o f = (uf)|Y_o$ for any $f \in L_b$;

d) the restriction of u_o to $L^1_{loc}(\mu) \cap \mathbb{R}^X$ is linear ;

e) we have $u_o(\bigvee_{n \in \mathbb{N}} f_n) = \bigvee_{n \in \mathbb{N}} (u_o f_n)$ for any upper bounded sequence $(f_n)_{n \in \mathbb{N}}$ in $L^1_{loc}(\mu)$;

f) for any $\nu \in M_b$ we have

$$\nu \in M_\mu \Longleftrightarrow \text{Supp } (v\nu) \subset Y_o \ ;$$

g) for any $f \in L^1_{loc}(\mu)$ we have

$$f \in L^1(\mu) \Longleftrightarrow u_o f \in L^1(v_o \mu) \Longrightarrow \int f d\mu = \int (u_o f) d(v_o \mu) \ .$$

We set

$$U_A := \text{Supp } v(1_A \cdot \mu)$$

for any $A \in \underline{\underline{R}}$. U_A is an open and compact set of Y ([9] Proposition 2.3). We set $Y_o := \bigcup_{A \in \underline{\underline{R}}} U_A$. Then Y_o is an open set of Y.

Let $A, B \in \underline{\underline{R}}$, $A \subset B$. Then

$$1_A \cdot \mu = 1_A \cdot (1_B \cdot \mu)$$

and therefore (Definition 2.1.1 d))

$$U_A = \text{Supp } (u1_A) \cdot v(1_B \cdot \mu) = (\text{Supp}(u1_A)) \cap U_B \subset \text{Supp } (u1_A)$$

We get $Y_o \subset \bigcup_{A \in \underline{\underline{R}}} \text{Supp } (u1_A)$ and

$$Y_o \cap \text{Supp}(u1_A) = U_A$$

for any $A \in \underline{\underline{R}}$. Hence Y_o is an open and closed set of $\bigcup_{A \in \underline{\underline{R}}} \mathrm{Supp}(u1_A)$.

Let $\nu \in M_\mu$. We have (Definition 2.1.1 d))

$$1_{U_A} \cdot v(1_B \cdot \nu) = (1_{U_A}(u1_A)) \cdot v(1_B \cdot \nu) = 1_{U_A} v((1_A 1_B) \cdot \nu) = 1_{U_A} v(1_A \cdot \nu)$$

and therefore

$$v(1_B \cdot \nu)|U_A = v(1_A \cdot \nu)|U_A$$

for any $A,B \in \underline{\underline{R}}$ with $A \subset B$. Since $\mathrm{Supp}\, v(1_A \cdot \nu) \subset U_A$ for any $A \in \underline{\underline{R}}$ it follows that there exists a unique $v_o \nu \in M(Y_o)$ such that

$$(v_o \nu)|U_A = v(1_A \cdot \nu)|U_A$$

for any $A \in \underline{\underline{R}}$. If $\nu \in M_\mu \cap M_b$ then (Definition 2.1.1. d))

$$(v\nu)|U_A = ((u1_A) \cdot (v\nu))|U_A = v(1_A \cdot \nu)|U_A = (v_o \nu)|U_A$$

for any $A \in \underline{\underline{R}}$ and therefore $(v\nu)|Y_o = v_o \nu$.

It is obvious that the map $v_o : M_\mu \longrightarrow M(Y_o)$ is linear and injective. Let $\lambda \in M(Y_o)_+$. For any $A \in \underline{\underline{R}}$ we denote by λ_A the unique measure of $M(Y)$ such that $\mathrm{Supp}\, \lambda_A \subset U_A, \lambda_A|U_A = \lambda|U_A$ and set $\nu_A := v^{-1}\lambda_A$. For any $A,B \in \underline{\underline{R}}$ with $A \subset B$ we have $\nu_A \leq \nu_B$ and (Definition 2.1.1 d))

$$\nu_B(A) = \int 1_A d\nu_B = \int (u1_A) d(v\nu_B) = \int (u1_A) d\lambda_B = \int (u1_A) d\lambda_A = \nu_A(A) .$$

Since M is a band of $M(\underline{R})$ there exists a unique $\nu \in M_+$ such that $1_A \cdot \nu = \nu_A$ for any $A \in \underline{\underline{R}}$. We have

$$(v_o \nu)|U_A = v(1_A \cdot \nu)|U_A = (v\nu_A)|U_A = \lambda_A|U_A = \lambda|U_A$$

for any $A \in \underline{\underline{R}}$ and therefore $v_o \nu = \lambda$. It follows that v_o is bijective. Since v is an isomorphism of vector lattices we deduce that v_o is an isomorphism of vector lattices.

The existence of u_o follows immediately from Radon-Nikodym theorem.

a) follows from the definition of Y_o .

b) we have

$$\mathrm{Supp}(v_o\mu) \supset \mathrm{Supp}\, v_o(1_A\mu) = \mathrm{Supp}(v(1_A\cdot\mu)|Y_o) = \mathrm{Supp}\, v(1_A\cdot\mu)$$

for any $A \in \underline{\underline{R}}$ and therefore

$$Y_o \subset \mathrm{Supp}(v_o\mu) \ .$$

c) Let $f \in L_b$ and let $A \in \underline{\underline{R}}$. We have (Definition 2.1.1 d))

$$((uf)|Y_o)\cdot v_o(1_A\cdot\mu) = ((uf)|Y_o)(v(1_A\cdot\mu)|Y_o) = (uf)\cdot v(1_A\cdot\mu)|Y_o =$$

$$= v((f1_A)\cdot\mu)|Y_o = v_o((f1_A)\cdot\mu) = (u_of)\cdot v_o(1_A\cdot\mu)$$

and therefore $(uf)|Y_o = u_of$ on U_A . Since A is arbitrary we get $(uf)|Y_o = u_of$.

d) and e) follow immediately from b) and Definition 2.2.1 b), c).

f) Let $\nu \in M_\mu \cap M_b$. Then

$$\mathrm{Supp}\, v(1_A\cdot\nu) \subset U_A \subset Y_o$$

for any $A \in \underline{\underline{R}}$ and therefore

$$\mathrm{Supp}(v\nu) = \overline{\bigcup_{A \in \underline{\underline{R}}} \mathrm{Supp}\, v\ (1_A\cdot\nu)} \subset \overline{Y}_o \ .$$

Let now $\nu \in M_b$ such that $\mathrm{Supp}\,(v\nu) \subset \overline{Y}_o$. Then $v(1_A\cdot\nu) << v(1_A\cdot\mu)$ and therefore $1_A\cdot\nu << 1_A\cdot\mu << \mu$ for any $A \in \underline{\underline{R}}$. We get $\nu << \mu$.

g) Let $f \in L^1(\mu)$. Then $f\cdot\mu \in M_\mu \cap M_b$ and by Definition 2.2.1 b), c)

$$(u_of)\cdot(v_o\mu) = v_o(f\cdot\mu) = v(f\cdot\mu)|Y_o \ .$$

Hence $u_of \in L^1(v_o\mu)$ and by f)

$$\int(u_of)d(v_o\mu) = \int 1_{Y_o} d(v(f\cdot\mu)|Y_o) = \int 1_Y dv(f\cdot\mu) = \int f d\mu \ .$$

Assume now $u_of \in L^1(v_o\mu)$. By Definition 2.2.1 c) we get

$$v_o(f\cdot\mu) = (u_of)\cdot(v_o\mu) \in M_b(Y_o) \ .$$

By f) and Definition 2.2.1 c) there exists $\nu \in M_\mu \cap M_b$ with

$$v_o \nu = v_o(f \cdot \mu)$$

and therefore $f \cdot \mu = \nu \in M_b$. Hence $f \in L^1(\mu)$.

Let (Y_1, u_1, v_1) be an arbitrary representation of μ associated to (Y,u,v) . We have

$$\operatorname{Supp} v_1(1_A \cdot \mu) = \operatorname{Supp} (v(1_A \cdot \mu)|Y_1) \subset \operatorname{Supp} v(1_A \cdot \mu) \subset Y_o$$

for any $A \in \underline{\underline{R}}$ and therefore

$$Y_1 = \operatorname{Supp} v_1(\mu) = \overline{\bigcup_{A \in \underline{\underline{R}}} \operatorname{Supp} v_1(1_A \cdot \mu)} \subset Y_o \ .$$

Let $\nu \in M_\mu \cap M_b$ such that $v_o \nu \leqslant 1_{Y_o \setminus Y_1}(v_o \mu)$. (Definition 2.2.1 a)) . Then

$$v_1 \nu = (v\nu)|Y_1 = ((v\nu)|Y_o)|Y_1 = (v_o \nu)|Y_1 = 0$$

and we get succesively

$$\nu = 0 \ , \ 1_{Y_o \setminus Y_1} \cdot (v_o \mu) = 0 \ , \ Y_o \setminus Y_1 = \emptyset \ , \ Y_o = Y_1 \ .$$

From this we deduce easily $v_1 = v_o$, $u_1 = u_o$. □

<u>*Proposition 2.2.3*</u> *Let (Y,u,v) be a bounded representation of (X,M), let $\mu, \nu \in M$ such that $\nu \ll \mu$, and let (Y_μ, u_μ, v_μ), (Y_ν, u_ν, v_ν) be representations of μ and ν respectively associated to (Y,u,v) . Then $\operatorname{Supp}(v_\mu \nu) = Y_\nu \subset Y_\mu$ and $(v_\mu \nu)|Y_\nu = v_\nu \nu$.*

For any $A \in \underline{\underline{R}}$ we have $1_A \cdot \nu \ll 1_A \cdot \mu$ and therefore $v(1_A \cdot \nu) \ll v(1_A \cdot \mu)$ (Definition 2.1.1 c)) and

$$\operatorname{Supp} v(1_A \cdot \nu) \subset \operatorname{Supp} v(1_A \cdot \mu) .$$

Hence (Proposition 2.2.2 a), b)) $Y_\nu \subset Y_\mu$, $\operatorname{Supp}(v_\mu \nu) \subset Y_\nu$.

For any $A \in \underline{\underline{R}}$ we have (Definition 2.2.1 b), c) Proposition 2.2.2 c))

$$((u1_A)|Y_\nu)\cdot((v_\mu\nu)|Y_\nu) = (((u1_A)|Y_\mu)\cdot(v_\mu\nu))|Y_\nu =$$
$$= ((u_\mu 1_A)\cdot(v_\mu\nu))|Y_\nu = v_\mu(1_A\cdot\nu)|Y_\nu = v(1_A\cdot\nu)|Y_\nu =$$
$$= v_\nu(1_A\cdot\nu) = u_\nu(1_A)\cdot(v_\nu\nu) = ((u1_A)|Y_\nu)\cdot v_\nu(\nu)$$

and therefore $(v_\mu\nu)|Y_\nu = v_\nu\nu$. We get further $\mathrm{Supp}\,(v_\mu\nu) = Y_\nu$. □

3. Representations of (X,M)

Definition 2.3.1 *An ordered triple* (Y,u,v) *is called a* *representation of* (X,M) *if* :

a) Y *is a locally compact hyperstonian space* ;

b) u *is a homomorphism of unital algebras* $L_\infty \longrightarrow C_\infty(Y)$;

c) $u(\bigvee_{n\in\mathbb{N}} f_n) = \bigvee_{n\in\mathbb{N}} (uf_n)$ *for any upper bounded sequence* $(f_n)_{n\in\mathbb{N}}$ *in* L_∞;

d) $u1_A$ *has a compact support for any* $A\in\underline{\underline{R}}$ *and*

$$Y = \bigcup_{A\in\underline{\underline{R}}} \mathrm{Supp}(u1_A) ;$$

e) v *is an isomorphism of vector lattices* $M \longrightarrow M(Y)$;

f) *we have*

$$f\in L^1(\mu) \Longleftrightarrow uf\in L^1(v\mu) \Longrightarrow \int f d\mu = \int (uf)d(v\mu) ,$$

$$f\in L^1_{\mathrm{loc}}(\mu) \Longleftrightarrow uf\in L^1_{\mathrm{loc}}(v\mu) \Longrightarrow v(f\cdot\mu) = (uf)\cdot(v\mu)$$

for any $(f,\mu)\in L_\infty\times M$.

Proposition 2.3.2 *Let* (Y,u,v) *be a representation of* (X,M) . *Then:*

a) $u^{-1}(C_i(Y)) = L,\ u^{-1}(C_b(Y)) = L_b,\ u^{-1}(C(Y)) = L_c$;

b) $v^{-1}(M_b(Y)) = M_b\ ,\ v^{-1}(M_c(Y)) = M_c$.

a) $u^{-1}(C_i(Y)) = L$ follows from Definition 2.3.1 e) and f) .

$u^{-1}(C_b(Y)) = L_b$ follows from Definition 2.3.1 b), c) and f) .

Let $f\in L_c$. Then $f1_A\in L$ for any $A\in\underline{\underline{R}}$; by the above result and Definition 2.3.1 b) $(uf)(u1_A)\in C_i(Y)$. In particular uf is bounded on $\mathrm{Supp}\,(u1_A)$ for any $A\in\underline{\underline{R}}$. By Definition 2.3.1 d) $uf\in C(Y)$.

Let $f\in L_\infty$ such that $uf\in C(Y)$. By Definition 2.3.1 b), d) $u(f1_A)\in C_b(Y)$ and by the above remark $f1_A\in L_b$ for any $A\in\underline{\underline{R}}$. Hence $f\in L_c$.

b) $v^{-1}(M_b(Y)) = M_b$ follows from Definition 2.3.1 b), f).

$v^{-1}(M_c(Y)) = M_c$ follows from Definition 2.3.1 d), f). □

Definition 2.3.3 *A representation* (Y,u,v) *of* (X,M) *and a bounded representation* (Y',u',v') *of* (X,M) *are called* associated *if* :

a) Y *is an open dense set of* Y' ;

b) $uf = (u'f)|Y$ *for any* $f \in L_b$;

c) $v\mu = (v'\mu)|Y$ *for any* $\mu \in M_b$.

By a) Y' is the Stone-Čech compactification of Y .

Proposition 2.3.4 *Any representation of* (X,M) *possesses a unique associated bounded representation of* (X,M) .

Let (Y,u,v) be a representation of (X,M) and let Y' be the Stone-Čech compactification of Y . Then $Y' \setminus Y$ is nowhere dense and therefore Y' is a compact hyperstonian space ([27] Exercise 23.2.11 C). By Proposition 2.3.2 a) $uf \in C_b(Y)$ for any $f \in C_b$; hence there exists a unique $u'f \in C_b(Y')$ whose restriction to Y equals uf . By Proposition 2.3.2 b) $v\mu \in M_b(Y)$ for any $\mu \in M_b$; hence there exists a unique $v'\mu \in M_b(Y')$ whose restriction to Y equals $v\mu$. It is easy to check that (Y',u',v') is a bounded representation of (X,M) . The uniqueness is obvious. □

Proposition 2.3.5 *Let* (Y,u,v) *be a representation of* (X,M) *and let* (Y',u',v') *be a bounded representation of* (X,M) . *Then there exists a unique continuous map* $\varphi: Y \longrightarrow Y'$ *such that* $v'\mu = \varphi(v\mu)$ *for any* $\mu \in M_b$. $\varphi(Y)$ *is the open dense set* $\bigcup_{A \in \underline{\underline{R}}} \text{Supp}\,(u'1_A)$ *of* Y', *the map* $Y \longrightarrow \varphi(Y)$ *defined by* φ *is a homeomorphism, and we have* $uf = (u'f)\circ\varphi$ *for any* $f \in L_b$.

Let (Y^*, u^*, v^*) be the bounded representation of (X,M) associated to (Y,u,v) (Proposition 2.3.4). By Theorem 2.1.2 there exists a homeomorphism $\varphi^* : Y^* \longrightarrow Y'$ such that $u^*f = (u'f)\circ\varphi^*$ for any $f \in L_b$ and $v'\mu = \varphi^*(v^*\mu)$ for any $\mu \in M_b$. We denote by φ the map $Y \longrightarrow Y'$ defined by φ^* . Then the map $Y \longrightarrow \varphi(Y)$ defined by φ^* is a homeomorphism and we have

$$uf = (u^*f)|Y = ((u'f)\circ\varphi^*)|Y = (u'f)\circ\varphi$$

for any $f \in L_b$. From this last relation and Definition 2.3.1 e) we get

$$\varphi(Y) = \varphi(\bigcup_{A\in \underline{\underline{R}}} \mathrm{Supp}(u1_A)) = \bigcup_{A\in \underline{\underline{R}}} \varphi(\mathrm{Supp}(u1_A)) =$$

$$= \bigcup_{A\in \underline{\underline{R}}} \varphi(\mathrm{Supp}((u'1_A)\circ\varphi)) = \bigcup_{A\in \underline{\underline{R}}} \mathrm{Supp}(u'1_A) .$$

It is obvious that $\bigcup_{A\in \underline{\underline{R}}} \mathrm{Supp}(u'1_A)$ is an open dense set of Y'. For any $\mu\in M_b$, $v\mu$ is the restriction of $v^*\mu$ to Y and therefore

$$\varphi(v\mu) = \varphi^*(v^*\mu) = v'\mu .$$

The unicity of φ is obvious. □

Theorem 2.3.6 Let (Y,u,v), (Y',u',v') be representations of (X,M). Then there exists a unique homeomorphism $\varphi: Y \longrightarrow Y'$ such that $v'\mu = \varphi(v\mu)$ for any $\mu\in M_c$. We have $uf = (u'f)\circ\varphi$ for any $f\in L_\infty$ and $v'\mu = \varphi(v\mu)$ for any $\mu\in M$.

Let (Y^*,u^*,v^*) be the bounded representation of (X,M) associated to (Y',u',v') (Proposition 2.3.4). By Proposition 2.3.5 there exists a continuous map $\varphi^*: Y \longrightarrow Y^*$ such that $\varphi^*(Y) = Y'$. The map $\varphi: Y \longrightarrow Y'$ defined by φ^* is a homeomorphism $uf = (u^*f)\circ\varphi^*$ for any $f \in L_b$, and $v^*\mu = \varphi^*(v\mu)$ for any $\mu\in M_b$. We get immediately $uf = (u'f)\circ\varphi$ for any $f\in L_b$ and $v'\mu = \varphi(v\mu)$ for any $\mu\in M_b$. We deduce

$$uf = u(\bigvee_{n\in\mathbb{N}} (f\wedge n1_X)) = \bigvee_{n\in\mathbb{N}} u(f\wedge n1_X) = \bigvee_{n\in\mathbb{N}} ((u'(f\wedge n1_X))\circ\varphi) =$$

$$= (\bigvee_{n\in\mathbb{N}} u'(f\wedge n1_X))\circ\varphi = (u'(\bigvee_{n\in\mathbb{N}} (f\wedge n1_X)))\circ\varphi = (u'f)\circ\varphi$$

for any $f\in L_{\infty+}$ and therefore $uf = (u'f)\circ\varphi$ for any $f\in L_\infty$. Similarly we deduce

$$v'\mu = v'(\bigvee_{A\in \underline{\underline{R}}} (1_A\cdot\mu)) = \bigvee_{A\in \underline{\underline{R}}} v'(1_A\cdot\mu) = \bigvee_{A\in \underline{\underline{R}}} \varphi(v(1_A\cdot\mu)) =$$

$$= \varphi(\bigvee_{A\in\underline{\underline{R}}} v(1_A\cdot\mu)) = \varphi(v(\bigvee_{A\in\underline{\underline{R}}}(1_A\cdot\mu))) = \varphi(v\mu)$$

for any $\mu\in M_+$ and therefore $v'\mu = \varphi(v\mu)$ for any $\mu\in M$.

The uniqueness of φ is obvious. □

Theorem 2.3.7 *Let* (Y,u,v) *be a bounded representation of* (X,M) . *Then there exists a unique representation* (Y',u',v') *of* (X,M) *associated to it and we have :*

a) $Y' = \bigcup_{A\in\underline{\underline{R}}} \mathrm{Supp}(u1_A)$;

b) *if* $\mu\in M$ *and if* (Y_μ,u_μ,v_μ) *denotes the representation of* μ *associated to* (Y,u,v) *then*

b_1) $Y_\mu \subset Y'$;

b_2) $u_\mu f = (u'f)|Y_\mu$ *for any* $f\in L_\infty \cap L^1_{loc}(\mu)$;

b)) $v_\mu \nu = (v'\nu)|Y_\mu$ and $\mathrm{Supp}(v_\mu\nu) = \mathrm{Supp}\,(v'\nu)$ *for any* $\nu\in M$ *with* $\nu \ll \mu$.

The uniqueness follows from Proposition 2.3.5.

We set

$$Y' := \bigcup_{A\in\underline{\underline{R}}} \mathrm{Supp}(u1_A)$$

and denote for any $\mu\in M$ by M_μ the band of M generated by μ and by (Y_μ,u_μ,v_μ) the representation of μ associated to (Y,u,v) (Proposition 2.2.2). We have

$$\sup_{A\in\underline{\underline{R}}} \int (u1_A)d\lambda = \sup_{A\in\underline{\underline{R}}} \int 1_A d(v^{-1}\lambda) = \int 1_X d(v^{-1}\lambda) = \int 1_Y d\lambda$$

for any $\lambda\in M_b(Y)_+$ and therefore Y' is an open dense set of Y which proves a) of Definition 2.3.3. In particular Y' is a hyperstonian space (N. Bourbaki, Topologie Générale Ch 1, § 11, ex 22 a). By Definition 2.2.1 a) Y_μ is an open and closed set of Y' for any $\mu\in M$ which proves b_1) .

Let $f\in L_{\infty+}$. We set

$$N := \{\mu\in M \mid f\in L^1_{loc}(\mu)\}.$$

By the definition of L_∞, N is a fundamental solid subspace of M and therefore $\bigcup_{\mu \in N} Y_\mu$ is an open dense set of Y'. By Proposition 2.2.2 c) we have

$$u(f \wedge n1_X)|Y_\mu = u_\mu(f \wedge n1_X)$$

for any $(n,\mu) \in \mathbb{N} \times N$ and therefore by Proposition 2.2.2 e)

$$\bigvee_{n \in \mathbb{N}} (u(f \wedge n1_X)|Y_\mu) = u_\mu(f)$$

for any $\mu \in N$. Hence the function

$$Y' \longrightarrow \bar{\mathbb{R}}, \ y \longmapsto \sup_{n \in \mathbb{N}} (u(f \wedge n1_X))(y)$$

is locally upper bounded (in $\mathbb{R}$) on a dense open set of Y'. We deduce that $\bigvee_{n \in \mathbb{N}} (u|f \wedge n1_X)|Y')$ exists in $C_\infty(Y')$; we denote it by $u'f$. For an arbitrary $f \in L_\infty$ we set

$$u'f := u'(f \vee 0) - u'((-f) \vee 0) .$$

$b_2)$ as well as b) of Definition 2.3.3. are obvious.

By Definition 2.2.1 a) for any $\mu \in M$ there exists a unique $v'\mu \in M(Y')$ such that

$$\mathrm{Supp}(v'\mu) = Y_\mu , \quad (v'\mu)|Y_\mu = v_\mu \mu .$$

In order to prove $b_3)$ let $\mu, \nu \in M$ with $\nu << \mu$. By Proposition 2.2.3

$$\mathrm{Supp}(v_\mu \nu) = Y_\nu \subset Y_\mu , \quad (v_\mu \nu)|Y_\nu = v_\nu \nu .$$

We get

$$\mathrm{Supp}(v_\mu \nu) = \mathrm{Supp}\,(v'\nu) , \ v_\mu \nu = (v'\nu)|Y_\mu .$$

We prove now that (Y',u',v') is a representation of (X,M).

Let $f,g \in L_{\infty+}$. We have

$$u'(f \wedge n1_X)+u'(g \wedge n1_X) = u(f \wedge n1_X)|Y'+u(g \wedge n1_X)|Y' =$$

$$= u(f \wedge n1_X+g \wedge n1_X)|Y' \leqslant u((f+g) \wedge 2n1_X)|Y' \leqslant$$

$$\leqslant u(f \wedge 2n1_X)|Y'+u(g \wedge 2n1_X)|Y' = u'(f \wedge 2n1_X)+u'(g \wedge 2n1_X) ,$$

$$u'(f\wedge n1_X)u'(g\wedge n1_X) = (u(f\wedge n1_X)|Y')(u(g\wedge n1_X))|Y' =$$

$$= u((f\wedge n1_X)(g\wedge n1_X))|Y' \leqslant u(fg\wedge n^2 1_X)\ Y' \leqslant$$

$$\leqslant u((f\wedge n^2 1_X)(g\wedge n^2 1_X))|Y' = u'(f\wedge n^2 1_X)u'(g\wedge n^2 1_X)$$

for any $n\in\mathbb{N}$ and therefore

$$u'(f) + u'(g) \leqslant u'(f+g) \leqslant u'(f) + u'(g) ,$$

$$u'(f)u'(g) \leqslant u'(fg) \leqslant u'(f)\,u'(g) .$$

From these relation we easily show that u' is a homomorphism of unital algebras.

By b_3) v' is an injective homomorphism of vector lattices. Let $\lambda\in M(Y')_+$ and let $A\in\underline{\underline{R}}$. Then there exists a unique $\mu_A\in M_b$ such that

$$(v\mu_A)|Y' = (u'1_A)\cdot\lambda .$$

We have $0 \leqslant \mu_A \leqslant \mu_B$ for any $A,B\in\underline{\underline{R}}$ with $A\subset B$ and therefore $(\mu_A)_{A\in\underline{\underline{R}}}$ is upper directed. We have

$$\mu_A(B) = \int 1_B d\mu_A = \int (u1_B)d(v\mu_A) = \int (u1_B)|Y'd(v\mu_A)|Y' =$$

$$= \int (u'1_B)d((u'1_A)\cdot\lambda) = \int (u'1_B)(u'1_A)d\lambda = \int (u'1_B)d\lambda$$

for any $A, B\in\underline{\underline{R}}$ with $A\supset B$ and therefore

$$\sup_{A\in\underline{\underline{R}}} \mu_A(B) = \int (u'1_B)d\lambda < \infty .$$

Hence $\bigvee_{A\in\underline{\underline{R}}} \mu_A$ exists in M ; we denote it by μ . By b_3) and by Definition 2.2.1 b) $v'\mu_A \leqslant v'\mu$ for any $A\in\underline{\underline{R}}$. Hence by a) and Definition 2.3.3 b)

$$v'\mu \geqslant \bigvee_{A\in\underline{\underline{R}}} (v'\mu_A) = \bigvee_{A\in\underline{\underline{R}}} ((v\mu_A)|Y') = \bigvee_{A\in\underline{\underline{R}}} ((u'1_A)\cdot\lambda) = \lambda .$$

We deduce $\operatorname{Supp}\lambda\subset Y_\mu$ and by b_3) and Definition 2.2.1 b) $\lambda\in v'(M)$. Hence v' is surjective and therefore an isomorphism of vector lattices.

Let $(f_n)_{n\in\mathbb{N}}$ be an upper bounded sequence in $\mathcal{L}_\infty$. By the definition of u' it follows immediately

$$u'f = \bigvee_{n\in\mathbb{N}} (u'f_n)$$

if all $f_n (n\in\mathbb{N})$ are positive. The general case may be reduced to this case by using the fact that u' is linear.

Let $(f,\mu)\in\mathcal{L}_\infty\times\mathcal{M}$. Assume $f\in\mathcal{L}^1(\mu)$. By Proposition 2.2.2 g) $u_\mu f\in\mathcal{L}^1(v_\mu\mu)$ and

$$\int f d\mu = \int (u_\mu f) d(v_\mu \mu) .$$

By $b_2)$ $(u'f)|Y_\mu = u_\mu f$ and by $b_3)$

$$v_\mu \mu = (v'\mu)|Y_\mu \ , \quad \mathrm{Supp}\,(v_\mu\mu) = \mathrm{Supp}\,(v'\mu) .$$

Hence $u'f\in\mathcal{L}^1(v'\mu)$ and

$$\int (u'f) d(v'\mu) = \int (u_\mu f) d(v_\mu\mu) = \int f d\mu .$$

Assume $f\in\mathcal{L}^1_{loc}(\mu)$. By the above results we get

$$(u'f)(u'1_A) = u'(f1_A)\in\mathcal{L}^1(v'\mu)$$

for any $A\in\underline{\underline{R}}$ and therefore $u'f\in\mathcal{L}^1_{loc}(v'\mu)$. We have

$$v'^{-1}((u'f)\cdot(v'\mu))(A) = \int 1_A dv'^{-1}((u'f)\cdot(v'\mu)) = \int (u'1_A) d((u'f)\cdot(v'\mu)) =$$

$$= \int (u'1_A)(u'f) d(v'\mu) = \int u'(f1_A) d(v'\mu) = \int f1_A d\mu = (f\cdot\mu)(A)$$

for any $A\in\underline{\underline{R}}$ and therefore

$$(u'f)\cdot(v'\mu) = v'(f\cdot\mu) .$$

Assume now f positive and $u'f\in\mathcal{L}^1(v'\mu)$. Then for any $(n,A)\in\mathbb{N}\times\underline{\underline{R}}$ we have $f\wedge n1_A\in\mathcal{L}^1(\mu)$ and by the above considerations we get

$$\int (f\wedge n1_A) d\mu = \int u'(f\wedge n1_A) d(v'\mu) .$$

We deduce

$$\sup_{(n,A)\in\mathbb{N}\times\underline{\underline{R}}} \int (f\wedge n1_A) d\mu = \int (u'f) d(v'\mu)$$

and therefore $f \in L^1(\mu)$. It follows at once that for an arbitrary f

$$u'f \in L^1(v'\mu) \implies f \in L^1(\mu) .$$

Assume now $u'f \in L^1_{loc}(v'\mu)$. Then $u'(f \wedge 1_A) \in L^1(v'\mu)$ and by the above result $f \wedge 1_A \in L^1(\mu)$ for any $A \in \underline{\underline{R}}$. Hence $f \in L^1_{loc}(\mu)$. □

Theorem 2.3.8 *There exists a representation and an associated bounded representation of* (X,M) .

The assertion follows immediately from Theorems 2.1.3 and 2.3.7. □

Proposition 2.3.9 *Let* Y *be a locally compact hyperstonian space, let* $L_\infty(Y)$ *be the set of real functions* f *on* Y *for which* $\{\lambda \in M(Y) \mid f \in L^1(\lambda)\}$ *is a fundamental solid subspace of* M(Y) *and let* v *be the identity map* $M(Y) \longrightarrow M(Y)$. *Then :*

a) for any $f \in L_\infty(Y)$ *there exists a unique element of* $C_\infty(Y)$, *which will be denoted by* uf , *such that* $\{f \neq uf\}$ *is nowhere dense ;*

b) (Y,u,v) *is a representation of* (Y,M(Y)) ;

c) u *is surjective .*

a) follows from [9] Proposition 3.6.
b) follows from [9] Proposition 3.5 and its Corollary.
c) follows from Proposition 1.5.1 and [9] Proposition 3.5. □

Remark It follows from the above result that any locally compact hyperstonian space appears in a representation of a measure space.

4. Supplementary results concerning the representations

Proposition 2.4.1 *Let* (Y,u,v) *be a representation of* (X,M) *and let* $(f_n)_{n \in \mathbb{N}}$ *be a sequence in* L_∞ . *Then* $(uf_n)_{n \in \mathbb{N}}$ *is upper bounded in* $C_\infty(Y)$ *iff* $(f_n)_{n \in \mathbb{N}}$ *is upper bounded in*

$$\{f \in \bar{\mathbb{R}}^X \mid M(f) \text{ is a fundamental solid subspace of } M\}$$

and in this case $f \in L_\infty$ *and*

$$uf = \bigvee_{n \in \mathbb{N}} (uf_n)$$

for any real function f *on* X *such that*

$$\sup_{n\in\mathbb{N}} f_n(x) < \infty \Longrightarrow f(x) = \sup_{n\in\mathbb{N}} f_n(x)$$

for any $x \in X$.

We denote by g the map

$$X \longrightarrow \bar{\mathbb{R}} \, , \quad x \longmapsto \sup_{n\in\mathbb{N}} f_n(x) \, .$$

Assume first $(uf_n)_{n\in\mathbb{N}}$ is upper bounded in $C_\infty(Y)$ and let N be the set of $\lambda \in M(Y)$ such that $\bigvee_{n\in\mathbb{N}} uf_n \in L^1(\lambda)$. Then N is a fundamental solid subspace of $M(Y)$ and by Definition 2.3.1 e) $v^{-1}(N)$ is a fundamental solid subspace of M . By Definition 2.3.1 c), f) $v^{-1}(N) \subset M(g)$, and therefore $M(g)$ is a fundamental solid subspace of M . It follows that $(f_n)_{n\in\mathbb{N}}$ is upper bounded in

$$\{f \in \bar{\mathbb{R}}^X \mid M(f) \text{ is a fundamental solid subspace of } M\} \, .$$

Let f be a real function on X such that

$$\sup_{n\in\mathbb{N}} f_n(x) < \infty \Longrightarrow f(x) = \sup_{n\in\mathbb{N}} f_n(x)$$

for any $x \in X$. By Proposition 1.5.1 $f \in L_\infty$ and $\{|g| = \infty\}$ is a μ-null set for any $\mu \in M$. For any $n \in \mathbb{N}$ we denote by f'_n the real function on X equal to f_n on $\{|g| < \infty\}$ and equal to f elsewhere. Then $f'_n \in L_\infty$ and by Definition 2.3.1 f) $uf_n = uf'_n$ for any $n \in \mathbb{N}$. From $f = \bigvee_{n\in\mathbb{N}} f'_n$ we get by Definition 2.3.1 c)

$$uf = \bigvee_{n\in\mathbb{N}} (uf'_n) = \bigvee_{n\in\mathbb{N}} (uf_n) \, .$$

Assume now that $(f_n)_{n\in\mathbb{N}}$ is upper bounded in

$$\{f \in \bar{\mathbb{R}}^X \mid M(f) \text{ is a fundamental solid subspace of } M\} .$$

Then $M(g)$ is a fundamental solid subspace of M . Let f be a real function on X equal to g on $\{|g| < \infty\}$. By Proposition 1.5.1 $f \in L_\infty$ and $f_n \leqslant f$ μ-a.e. for any $\mu \in M$. By Definition 2.3.1 f) we get $uf_n \leq uf$ for any $n \in \mathbb{N}$. Hence $(uf_n)_{n\in\mathbb{N}}$ is upper bounded in $C_\infty(Y)$. □

Proposition 2.4.2 *Let* (Y,u,v) *be a representation of* (X,M). *Then for any* $f \in L_\infty$ *and for any Borel set* B *of* $\bar{\mathbb{R}}$ *the set*

$$\{u1^X_{f^{-1}(B)} \neq 1^Y_{\widehat{uf}^{-1}(B)}\}$$

is nowhere dense.

Let $\underline{\underline{B}}$ be the set of Borel sets of $\bar{\mathbb{R}}$ for which

$$\{ u1^X_{f^{-1}(B)} \neq 1^Y_{\widehat{uf}^{-1}(B)}\}$$

is nowhere dense for any $f \in L_\infty$.

Let $f \in L_\infty$. We set $A:=\{f>0\}$, $A':=\{uf>0\}$. Then $(1^X_X \wedge n(f \vee 0))_{n\in\mathbb{N}}$ is an increasing sequence in L_∞ whose supremum is 1^X_A. $(1^Y_Y \wedge n(uf \vee 0))_{n\in\mathbb{N}}$ is an increasing sequence and by Definition 2.3.1 a) c) its supremum in $C_\infty(Y)$ is $u1^X_A$. On the other side $1^Y_{A'}$ is the supremum in $\mathbb{R}^Y$ of this sequence and therefore the set $\{u1^X_A \neq 1^Y_{A'}\}$ is nowhere dense. Since f is arbitrary we deduce $]0,\infty[\in \underline{\underline{B}}$. From this we easily deduce $]\alpha,\infty]$, $[-\infty,\alpha[\in \underline{\underline{B}}$ for any $\alpha \in \mathbb{R}$. Since $\underline{\underline{B}}$ is a σ-algebra ([9] Corollary of the Proposition 5) it follows that it contains any Borel set of $\bar{\mathbb{R}}$. □

Proposition 2.4.3 *Let* I *be a countable set, let* φ *be a Borel measurable real function on* $\bar{\mathbb{R}}^I$, *and let* $(f_\iota)_{\iota\in I}$ *be a family in* L_∞. *For any set* Z *and for any family* $(g_\iota)_{\iota\in I}$ *in* $\bar{\mathbb{R}}^Z$ *we denote by* $\varphi((g_\iota)_{\iota\in I})$ *the map*

$$Z \longrightarrow \mathbb{R}\,,\; z \longmapsto \varphi((g_\iota(z))_{\iota\in I})\,.$$

Then:

a) $\varphi((f_\iota)_{\iota\in I}) \in \mathcal{L}_\infty$;

b) $\{u\varphi((f_\iota)_{\iota\in I}) \neq \varphi((uf_\iota)_{\iota\in I})\}$ *is a nowhere dense set for any representation* (Y,u,v) *of* (X,M) ;

c) $u\varphi((f_\iota)_{\iota\in I}) = u\psi((f_\iota)_{\iota\in I})$ *for any Borel measurable real function* ψ *on* $\bar{\mathbb{R}}^I$ *for which* $\psi|\mathbb{R}^I = \varphi|\mathbb{R}^I$.

a) & b) Let Φ be the set of Borel measurable real functions on $\bar{\mathbb{R}}^I$ possessing the above properties a) and b). Φ is a subspace of the space of real functions on $\bar{\mathbb{R}}^I$.

Let $(\varphi_n)_{n\in\mathbb{N}}$ be an increasing sequence in Φ such that $\sup \varphi_n(z) < \infty$ for any $z \in \bar{\mathbb{R}}^I$ and let φ be the map

$$\bar{\mathbb{R}}^I \longrightarrow \mathbb{R} \, , \; z \longmapsto \sup_{n\in\mathbb{N}} \varphi_n(z) \, ,$$

Then φ is a Borel measurable real function on $\bar{\mathbb{R}}^I$. Since I is countable the set

$$\bigcup_{n\in\mathbb{N}} \{u\varphi_n((f_\iota)_{\iota\in I}) \neq \varphi_n((uf_\iota)_{\iota\in I})\}$$

is meagre and therefore nowhere dense ([9] Corollary of Proposition 5). The function $\varphi((uf_\iota)_{\iota\in I})$ is lower semi-continuous outside this set and therefore it possesses a majorant in $C_\infty(Y)$. Hence $(u\varphi_n((f_\iota)_{\iota\in I}))_{n\in\mathbb{N}}$ is upper bounded in $C_\infty(Y)$ and by Proposition 2.4.1 c) we get $\varphi((f_\iota)_{\iota\in I}) \in \mathcal{L}_\infty$ and

$$\{u\varphi((f_\iota)_{\iota\in I}) \neq \varphi((uf_\iota)_{\iota\in I})\}$$

is a nowhere dense set. Hence $\varphi\in\Phi$.

Let $(B_\iota)_{\iota\in I}$ be a family of Borel sets of $\bar{\mathbb{R}}$ such that $\{\iota\in I \mid B_\iota \neq \bar{\mathbb{R}}\}$ is finite and let φ be the characteristic function of $\prod_{\iota\in I} B_\iota$. Then

$$\varphi((uf_\iota)_{\iota\in I}) = 1^Y_{\bigcap_{\iota\in I} \overline{uf}_\iota^{-1}(B_\iota)} = \bigwedge_{\iota\in I}^{\mathbb{R}^Y} 1^Y_{\overline{uf}_\iota^{-1}(B_\iota)} \, ,$$

$$\varphi((f_\iota)_{\iota\in I}) = 1^X_{\bigcap_{\iota\in I} \bar{f}_\iota^{-1}(B_\iota)} = \bigwedge_{\iota\in I} 1^X_{\bar{f}^{-1}(B_\iota)} \in \mathcal{L}_\infty$$

and therefore

$$u\varphi((f_\iota)_{\iota\in I}) = \bigwedge_{\iota\in I}(u\, 1^X_{\bar{f}^1_\iota(B_\iota)}) .$$

By Proposition 2.4.2

$$\{1^Y_{\widehat{uf_\iota^{-1}(B_\iota)}} \neq u\, 1^X_{\bar{f}^1_\iota(B_\iota)}\}$$

is nowhere dense for any $\iota\in I$ and therefore

$$\{u\varphi((f_\iota)_{\iota\in I} \neq \varphi((uf_\iota)_{\iota\in I})\}$$

is a nowhere dense set. Hence $\varphi\in\Phi$.

From the above considerations we deduce that any Borel measurable real function on $\bar{\mathbb{R}}^I$ belongs to Φ .

c) Let $\mu\in M$. By Proposition 1.5.1 $\{|f_\iota| = \infty\}$ is a μ-null set. Since I is countable $\bigcup_{\iota\in I}\{|f_\iota| = \infty\}$ is a μ-null set. Hence

$$\{\varphi((f_\iota)_{\iota\in I}) \neq \psi((f_\iota)_{\iota\in I})\}$$

is a μ-null set too. μ being arbitrary we get

$$u\varphi((f_\iota)_{\iota\in I}) = u\psi((f_\iota)_{\iota\in I}) . \square$$

Proposition 2.4.4 *Let* (Y,u,v) *be a representation of* (X,M), *let* (Y',u',v') *be the associated bounded representation of* (X,M) *and for any subset* M *of* Y' *let* $\bar{M}$ *be its closure in* Y' . *We set :*

$$M_\ell := \{\mu\in M \mid \mu \text{ possesses a locally countable concassage}\} ,$$

$$Y_c := \bigcup_{\mu\in M_c} \operatorname{Supp}(v\mu), Y_\ell := \bigcup_{\mu\in M_\ell} \overline{\operatorname{Supp}(v\mu)}, Y_m := \bigcup_{\mu\in M} \overline{\operatorname{Supp}(v\mu)} .$$

Then:

a) Y_c, Y_ℓ, Y_m *are open dense sets of* Y' *and* $Y_c\subset Y_\ell\subset Y_m$;

b) *if* N, N' *are countable subsets of* M *and* M_ℓ *respectively such that* $|\mu|\wedge|\mu'| = 0$ *for any* $(\mu,\mu')\in N\times N'$ *then there exists a measurable subset* A *of* X *such that* $u'1_A = 0$ *on* $\overline{\bigcup_{\mu\in N} \mathrm{Supp}(v\mu)}$ *and* $u'1_A = 1$ *on* $\overline{\bigcup_{\mu\in N'} \mathrm{Supp}(v\mu)}$;

c) *for any* $g\in C_b(Y')$ *and for any countable subset* N *of* M_ℓ *there exists* $f\in L_b$, f *measurable, such that* $u'f = g$ *on* $\overline{\bigcup_{\mu\in N} \mathrm{Supp}(v\mu)}$;

d) *for any* $g\in C_\infty(Y')$ *and for any countable subset* N *of* M_ℓ *there exists a measurable real function* f *on* X, *such that* $uf = g$ *on* $\bigcup_{\mu\in N} \mathrm{Supp}(v\mu)$;

e) *for any sequence* $(K_n)_{n\in\mathbb{N}}$ *of compact sets of* Y_c *we have*

$$\overline{(\bigcup_{n\in\mathbb{N}} K_n)} \cap Y \subset Y_c .$$

a) Let U be an open nonempty set of Y'. There exists $\mu\in M$ with $\mu>0$ and $\mathrm{Supp}(v\mu)\subset U$. We have $\mu = \bigvee_{A\in\underline{\underline{R}}} (1_A\cdot\mu)$ and therefore there exists $A\in\underline{\underline{R}}$ with $1_A\cdot\mu \neq 0$. Since $1_A\cdot\mu\in M_c$ we get

$$\emptyset \neq \mathrm{Supp}\ v(1_A\cdot\mu) \subset \mathrm{Supp}(v\mu)\cap Y_c \subset U\cap Y_c .$$

Hence Y_c is dense. The other assertions are obvious.

b) follows from Propositions 1.4.2 and 1.4.3 (and Proposition 1.5.2, Proposition 1.5.3 d)).

c) We set

$$Z := \overline{\bigcup_{\mu\in N} \mathrm{Supp}(v\mu)} .$$

Let $x,y\in Z$, $x\neq y$. Let U,V be open disjoint sets of Y' containing x and y respectively. We set

$$N' := \{v^{-1}(1_{U\cap Y}\cdot\mu)\,|\,\mu\in N\},\ N'' := \{v^{-1}(1_{V\cap Y}\cdot\mu)\,|\,\mu\in N\}.$$

Then N', N'' are countable subsets of M_ℓ such that $|\mu'|\wedge|\mu''| = 0$ for any $(\mu',\mu'')\in N'\times N''$. By b) there exists a measurable subset A of X such that

$$(u'1_A)(x) = 0 ,\quad (u'1_A)(y) = 1 .$$

By this result, by Proposition 1.5.3 b), and by Definition 2.3.1 b) we may apply Weierstrass-Stone theorem and get c).

d) We may assume g positive. Then by c) for any $n \in \mathbb{N}$ there exists a measurable $f_n \in \mathcal{L}_b$ such that $u'f_n = g \wedge n$ on $\bigcup_{\mu \in N} \mathrm{Supp}(v\mu)$. We have $\bigvee_{n \in \mathbb{N}} f_n \in \bigcap_{\mu \in N} \mathcal{L}^1_{loc}(\mu)$ and therefore $\bigvee_{n \in \mathbb{N}} f_n = \infty$ is a μ-null set for any $\mu \in N$. We denote by f the function on X equal to f on $\{\bigvee_{n \in \mathbb{N}} f_n < \infty\}$ and equal to 0 elsewhere. It is obvious that f is a real measurable function on X and belongs therefore to $\mathcal{L}_\infty$.

Let $\mu \in N$ and let (Y_μ, u_μ, v_μ) be the representation of μ associated to (Y', u', v'). By Definition 2.2.1 c) and Proposition 2.2.2 b), c), e) we have

$$u_\mu f = u_\mu(\bigvee_{n \in \mathbb{N}} f_n) = \bigvee_{n \in \mathbb{N}} (u_\mu f_n) = \bigvee_{n \in \mathbb{N}} ((u'f_n)|Y_\mu) = g|Y_\mu .$$

By Theorem 2.3.7 b) $(uf)|Y_\mu = u_\mu f$.

Since μ is arbitrary we get $uf = g$ on $\bigcup_{\mu \in N} \mathrm{Supp}(v\mu)$.

e) Let $x \in \overline{(\bigcup_{n \in \mathbb{N}} K_n)} \cap Y$. There exists $A \in \underline{\underline{R}}$ with $x \in \mathrm{Supp}(u1_A)$ (Theorem 2.3.7 a)). For any $n \in \mathbb{N}$ there exists $\mu_n \in M_c$ with $\|\mu_n\| = 1$ and

$$K_n \cap \mathrm{Supp}(u1_A) \subset \mathrm{Supp}\mu_n \subset \mathrm{Supp}(u1_A) .$$

Then

$$\mu := \sum_{n \in \mathbb{N}} \frac{1}{n^2} |\mu_n| \in M_c$$

and

$$x \in \overline{\bigcup_{n \in \mathbb{N}} (K_n \cap \mathrm{Supp}(u1_A))} \subset \overline{\bigcup_{n \in \mathbb{N}} \mathrm{Supp}(v\mu_n)} \subset \mathrm{Supp}(v\mu) \subset Y_c . \square$$

§ 3 DUALS OF SPACES OF MEASURES

1. Structure on M^ρ

For any $\xi \in M^\rho$ we denote by $M(\xi)$ the largest fundamental solid subspace of M on which ξ is defined. Let $\mu \in M$. We set

$$\hat{\mathcal{L}}^1(\mu) := \{\xi \in M^\rho \mid \mu \in M(\xi)\},$$

$$\hat{\mathcal{L}}^1_{\mathrm{loc}}(\mu) := \{\xi \in M^\rho \mid \forall A \in \underline{\underline{R}} \Longrightarrow 1_A \cdot \mu \in M(\xi)\} ,$$

and call the elements of $\hat{\mathcal{L}}^1(\mu)$ $(\hat{\mathcal{L}}^1_{\mathrm{loc}}(\mu))$ <u>μ-integrable</u> (<u>locally μ-integrable</u>). For any $\xi \in \hat{\mathcal{L}}^1(\mu)$ $(\hat{\mathcal{L}}^1_{\mathrm{loc}}(\mu))$ we set

$$\int \xi d\mu := \langle \mu, \xi \rangle ,$$

$$(\xi \cdot \mu : \underline{\underline{R}} \longrightarrow \mathbb{R} , A \longmapsto \langle 1_A \cdot \mu \; \xi \rangle) .$$

We have

$$f \in \mathcal{L}^1(\mu) \Longleftrightarrow \dot{f} \in \hat{\mathcal{L}}^1(\mu) \Longrightarrow \int f d\mu = \int \dot{f} d\mu$$

for any $f \in \mathcal{L}_\infty$. For any $\xi \in \hat{\mathcal{L}}^1_{\mathrm{loc}}(\mu)$ we have $\xi \cdot \mu \in M$. We set

$$\hat{\mathcal{L}}(\mu) := \{\xi \in \hat{\mathcal{L}}^1_{\mathrm{loc}}(\mu) \mid \xi \cdot \mu = 0\} ,$$

$$\hat{\mathcal{L}}^\perp(\mu) := \{\xi \in M^\rho \mid \forall \eta \in \hat{\mathcal{L}}(\mu) \Longrightarrow |\xi| \wedge |\eta| = 0\} .$$

<u>*Proposition 3.1.1*</u> *Let* (Y,u,v) *be a representation of* (X,M) *. Then:*

a) *for any* $\xi \in M^\rho$ *there exists a unique element of* $C_\infty(Y)$*, which will be denoted by* $\hat{u}\,\xi$*, such that*

$$\xi \in \hat{\mathcal{L}}^1(\mu) \Longleftrightarrow \hat{u}\xi \in \mathcal{L}^1(v\mu) \Longrightarrow \int \xi d\mu = \int \hat{u}\xi d(v\mu)$$

for any $\mu \in M$;

b) $\hat{u} ; M^\rho \longrightarrow C_\infty(Y)$ *is an isomorphism of vector lattices* ;

c) $\hat{u}(M^\pi_c) = C(Y)$,

$\hat{u}(M^\pi_b) = C_b(Y)$,

$\hat{u}(M^\pi) = C_i(Y)$;

d) $\hat{u}\dot{f} = uf$ *for any* $f \in L_\infty$;

e) *for any* $(\xi,\mu) \in M^\rho \times M$ *we have*

$$\xi \in \hat{L}^1_{loc}(\mu) \Longleftrightarrow \hat{u}\xi \in L^1_{loc}(v\mu) \Longrightarrow v(\xi\cdot\mu) = (u\xi)\cdot(v\mu) \quad .$$

Let $v' : M(Y)^\rho \longrightarrow M^\rho$ be the adjoint isomorphism of v and let w be the map

$$C_\infty(Y) \longrightarrow M^\rho, \; f \longmapsto v'\dot{f} \; .$$

By Proposition 1.6.1 a) w is an isomorphism of vector lattices. Its inverse map is $\hat{u}$ which proves a) and b). c) and d) are obvious.

e) The equivalence

$$\xi \in \hat{L}^1_{loc}(\mu) \Longleftrightarrow \hat{u}\xi \in L^1_{loc}(v\mu)$$

is obvious. For any $A \in \underline{\underline{R}}$ we have

$$(\xi\cdot\mu)(A) = \langle 1_A\cdot\mu,\xi\rangle = \int \xi d(1_A\cdot\mu) = \int \hat{u}\xi dv(1_A\cdot\mu) =$$

$$= \int \hat{u}\xi d((u1_A)(v\mu)) = \int u1_A d((\hat{u}\xi)\cdot(v\mu)) = \int 1_A dv^{-1}((\hat{u}\xi)\cdot(v\mu)) =$$

$$= v^{-1}((\hat{u}\xi)\cdot(v\mu))(A)$$

and therefore

$$v(\xi\cdot\mu) = (\hat{u}\xi)\cdot(v\mu) \; . \; \square$$

<u>*Corollary 3.1.2*</u> *We have for any* $\xi \in M^\rho$;

a) $|\xi| \wedge \dot{1}_X = 0 \Longleftrightarrow \xi = 0$;

b) $\xi \in M^\pi_b \Longrightarrow (\exists n \in \mathbb{N} \, , \, |\xi| \leq n\dot{1}_X)$

c) $\eta \in M^\rho, \; |\eta| \leq |\xi| \Longrightarrow \exists \in M^\pi_b \; , \; |\zeta| \leq \dot{1}_X \; , \; \eta = \xi\zeta$

The assertions follow immediately from Theorem 2.3.8 and Proposition 3.1.1 a), b), c), d) . $\square$

<u>*Corollary 3.1.3*</u> *For any* $\mu \in M$ *we have :*

a) $\hat{L}^1(\mu)$, $\hat{L}^1_{loc}(\mu)$ *are fundamental solid subspaces of* M^ρ ;

b) $\hat{L}^1(\mu) \subset \hat{L}^1_{loc}(\mu)$;

c) $\hat{L}(\mu)$ *is a band of* M^ρ .

The assertions follow immediately from Theorem 2.3.8 and Proposition 3.1.1 a), b), e) . □

Corollary 3.1.4 *Let* $\mu,\nu \in M$ *such that* $|\mu| \wedge |\nu| = 0$. *Then* $\hat{L}^\perp(\mu) \cap \hat{L}^\perp(\nu) = \{0\}$.

The assertion follows from Thoerem 2.3.8 and Proposition 3.1.1 e) .□

Theorem 3.1.5 *Let* I *be a countable set and let* φ *be a Borel measurable real function on* $\bar{\mathbb{R}}^I$. *For any set* Z *and for any family* $(g_\iota)_{\iota\in I}$ *in* $\bar{\mathbb{R}}^Z$ *we denote by* $\varphi((g_\iota)_{\iota\in I})$ *the map*

$$Z \longrightarrow \mathbb{R}\ ,\ z \longmapsto \varphi((g_\iota(z))_{\iota\in I})\ .$$

Then there exists a unique map $\hat{\varphi} : (M^\rho)^I \longrightarrow M^\rho$ *such that for any representation* (Y,u,v) *of* (X,M) *and for any family* $(\xi_\iota)_{\iota\in I}$ *in* M^ρ *the set*

$$\{\hat{u}\hat{\varphi}((\xi_\iota)_{\iota\in I}) \neq \varphi((\hat{u}\xi_\iota)_{\iota\in I})\}$$

is nowhere dense. For any family $(f_\iota)_{\iota\in I}$ *in* L_∞ *we have* $\varphi((f_\iota)_{\iota\in I}) \in L_\infty$ *and*

$$\hat{\varphi}((\dot{f}_\iota)_{\iota\in I}) = \dot{\widehat{\varphi((f_\iota)_{\iota\in I})}}\ .$$

Let $(\xi_\iota)_{\iota\in I}$ be a family in M^ρ . Let (Y,u,v) be a representation of (X,M) (Theorem 2.3.8) and let $L_\infty(Y)$ be the set of real functions f on Y which $\{\lambda\in M(Y)\,|\,f\in L^1(\lambda)\}$ is a fundamental solid subspace of $M(Y)$. By Proposition 3.1.1 a) and Proposition 2.4.3 a) $\varphi((\hat{u}\xi_\iota)_{\iota\in I}) \in L_\infty(Y)$. By Theorem 2.3.9 a) there exists a unique $f\in C_\infty(Y)$ such that $\{f \neq \varphi((u\xi_\iota)_{\iota\in I})\}$ is a nowhere dense set. By Proposition 2.3.9 b) and Proposition 3.1.1 b) there exists a unique $\xi\in M^\rho$ such that $\hat{u}\xi = f$. We want to show that ξ does not depend on the representation (Y,u,v) .

Let (Y',u',v') be another representation of (X,M) and let ξ' be the element of M^ρ for which $\{\hat{u}'\xi' \neq \varphi((\hat{u}'\xi_\iota')_{\iota\in I})\}$ is a nowhere dense set. By Theorem 2.3.6 there exists a homeomorphism $\psi = Y \longrightarrow Y'$ such that $v'\mu = \psi(v\mu)$ for any $\mu\in M$. By Proposition 1.6.1 a), b) and Proposition 3.1.1 a) we get $\hat{u}\eta = (\hat{u}'\eta)\circ\psi$ for any $\eta \in M^{\rho}$. We deduce $\hat{u}\xi' = (\hat{u}\xi')\circ\psi$,

$$\varphi((\hat{u}'\xi_\iota)_{\iota\in I})\circ\psi = \varphi(((\hat{u}'\xi_\iota)\circ\psi)_{\iota\in I}) = \varphi((\hat{u}\xi_\iota)_{\iota\in I})$$

and therefore

$$\{\hat{u}\xi' \neq \varphi((\hat{u}\xi_\iota)_{\iota\in I})\} = \{(\hat{u}'\xi')\circ\psi \neq \varphi((\hat{u}'\xi_\iota)_{\iota\in I})\circ\psi\} =$$

$$= \psi^{-1}(\{\hat{u}'\xi' \neq \varphi((\hat{u}'\xi_\iota)_{\iota\in I})\}.$$

Hence $\{\hat{u}\xi' \neq \varphi((\hat{u}\xi_\iota)_{\iota\in I})$ is nowhere dense and we get $\hat{u}\xi' = \hat{u}\xi$ and therefore $\xi = \xi'$.

Let now $(f_\iota)_{\iota\in I}$ be a family in L_∞. By Proposition 2.4.3 a) $\varphi((f_\iota)_{\iota\in I}) \in L_\infty$. Let (Y,u,v) be a representation of (X,M) (Theorem 2.3.8). The set

$$\{\hat{u}\hat{\varphi}((\dot{f}_\iota)_{\iota\in I}) \neq \varphi((\hat{u}\dot{f}_\iota)_{\iota\in I})\}$$

is nowhere dense. By Proposition 3.1.1 d) $\hat{u}\dot{f}_\iota = uf_\iota$ for any $\iota\in I$ and $\hat{u}\,\dot{\overbrace{\varphi((f_\iota)_{\iota\in I})}} = u\varphi((f_\iota)_{\iota\in I})$. By Proposition 2.4.3 b) the set

$$\{u\varphi((f_\iota)_{\iota\in I}) \neq \varphi((uf_\iota)_{\iota\in I})\}$$

is nowhere dense. Hence $\hat{u}\hat{\varphi}((\dot{f}_\iota)_{\iota\in I}) = \hat{u}\,\dot{\overbrace{\varphi((f_\iota)_{\iota\in I})}}$ and therefore $\hat{\varphi}((\dot{f}_\iota)_{\iota\in I}) = \dot{\overbrace{\varphi((f_\iota)_{\iota\in I})}}$. □

<u>Remark</u> Let I be the set $\{0,1\}$ and let φ be the map

$$\overline{\mathbb{R}}^2 \longrightarrow \mathbb{R}, \quad (\alpha,\beta) \longmapsto \begin{cases} \alpha+\beta & \text{if } (\alpha,\beta)\in \mathbb{R}^2 \\ 0 & \text{otherwise.} \end{cases}$$

Then $\hat{\varphi}(\xi,\eta) = \xi+\eta$ for any $(\xi,\eta)\in M^{\rho}\times M^{\rho}$.

<u>*Proposition 3.1.6*</u> *Let* I *be a countable set, let* $(I_\kappa)_{\kappa\in K}$ *be a disjoint family of sets such that* $I = \bigcup_{\kappa\in K} I_\kappa$, *let* φ *and* ψ *be Borel measurable real functions on* $\overline{\mathbb{R}}^I$ *and* $\overline{\mathbb{R}}^K$ *respectively, and for each* $\kappa\in K$ *let* φ_κ *be a Borel measurable real function on* $\overline{\mathbb{R}}^{I_\kappa}$ *such that*

$$\varphi((\alpha_\iota)_{\iota\in I}) = \psi((\varphi_\kappa((\alpha_\iota)_{\iota\in I_\kappa}))_{\kappa\in K})$$

for any $(\alpha_\iota)_{\iota\in I} \in \mathbb{R}^I$. *Then with the notations of Theorem 3.1.5*

$$\hat{\varphi}((\xi_\iota)_{\iota\in I}) = \hat{\psi}((\hat{\varphi}_\kappa((\xi_\iota)_{\iota\in I_\kappa}))_{\kappa\in K})$$

for any $(\xi_\iota)_{\iota\in I} \in (M^\rho)^I$.

Let (Y,u,v) be a representation of (X,M). The sets

$$\{\hat{u}\hat{\varphi}((\xi_\iota)_{\iota\in I}) \neq \varphi((\hat{u}\xi_\iota)_{\iota\in I})\} ,$$

$$\{\hat{u}\hat{\psi}((\hat{\varphi}_\kappa((\xi_\iota)_{\iota\in I_\kappa}))_{\kappa\in K}) \neq \psi((\hat{u}\hat{\varphi}_\kappa((\xi_\iota)_{\iota\in I_\kappa}))_{\kappa\in K})\} ,$$

$$\bigcup_{\kappa\in K}\{\hat{u}\hat{\varphi}_\kappa((\xi_\iota)_{\iota\in I_\kappa}) \neq \varphi_\kappa((\hat{u}\xi_\iota)_{\iota\in I_\kappa})\} ,$$

$$\bigcup_{\iota\in I}\{|\hat{u}\xi_\iota| = \infty\}$$

are nowhere dense ([9] Corollary of Proposition 3.5) and therefore

$$\{\hat{u}\hat{\varphi}((\xi_\iota)_{\iota\in I}) \neq \hat{u}\hat{\psi}((\hat{\varphi}_\kappa((\xi_\iota)_{\iota\in I_\kappa}))_{\kappa\in K})\}$$

is a nowhere dense set. We deduce

$$\hat{u}\hat{\varphi}((\xi_\iota)_{\iota\in I}) = \hat{u}\hat{\psi}((\hat{\varphi}_\kappa((\xi_\iota)_{\iota\in I_\kappa}))_{\kappa\in K})$$

and by Proposition 3.1.1 b)

$$\hat{\varphi}((\xi_\iota)_{\iota\in I}) = \hat{\psi}((\hat{\varphi}_\kappa((\xi_\iota)_{\iota\in I_\kappa}))_{\kappa\in K}) \quad . \square$$

<u>*Theorem 3.1.7*</u> *There exists a unique map*

$$M^\rho \times M^\rho \longrightarrow M^\rho , \quad (\xi,\eta) \longmapsto \xi\eta$$

such that :

1) $(\xi+\eta)\zeta = \xi\zeta + \eta\zeta$, $\zeta(\xi+\eta) = \zeta\xi + \zeta\eta$ *for any* $\xi,\eta,\zeta \in M^\rho$,

2) $\xi \dot{1}_X = \dot{1}_X \xi = \xi$ *for any* $\xi \in M_b^\pi$;

3) *for any* $\xi \in M_+^\rho$ *and for any upper bounded sequence* $(\xi_n)_{n\in\mathbb{N}}$ *in* M_+^ρ *for which the families* $(\xi\,\xi_n)_{n\in\mathbb{N}}$, $(\xi_n\,\xi)_{n\in\mathbb{N}}$ *are upper bounded we have*

$$\xi\left(\bigvee_{n\in\mathbb{N}} \xi_n\right) = \bigvee_{n\in\mathbb{N}} (\xi\,\xi_n) \ , \ \left(\bigvee_{n\in\mathbb{N}} \xi_n\right)\xi = \bigvee_{n\in\mathbb{N}} (\xi_n \xi) \ .$$

M^ρ *endowed with this composition law is a commutative, associative, unital algebra. We have:*

a) *for any* $\xi \in M_+^\rho$ *and for any upper bounded nonempty family* $(\xi_\iota)_{\iota\in I}$ *in* M^ρ *the family* $(\xi\,\xi_\iota)_{\iota\in I}$ *is upper bounded and*

$$\xi\left(\bigvee_{\iota\in I} \xi_\iota\right) = \bigvee_{\iota\in I} (\xi\,\xi_\iota) \ ;$$

b) *the map*

$$L_\infty \longrightarrow M^\rho \ , \quad f \longmapsto \dot{f}$$

is an homomorphism of unital algebras ;

c) M^π, M_b^π, M_c^π *are fundamental solid subspaces of* M^ρ *such that* $M^\pi \subset M_b^\pi \subset M_c^\pi$;

d) M_b^π *and* M_c^π *are unital subalgebras of* M^ρ ;

e) M^π *is an ideal of* M_c^π ;

f) *if* φ *denotes the Borel measurable real function on* $\bar{\mathbb{R}}^2$

$$\bar{\mathbb{R}}^2 \longrightarrow \bar{\mathbb{R}} \ , \quad (\alpha,\beta) \longmapsto \begin{cases} \alpha\beta & \text{if } (\alpha,\beta)\in\mathbb{R}^2 \\ 0 & \text{if } (\alpha,\beta)\in\bar{\mathbb{R}}^2\setminus\mathbb{R}^2 , \end{cases}$$

then $\xi\eta = \hat{\varphi}(\xi,\eta)$ *for any* $(\xi,\eta)\in M^\rho\times M^\rho$ *(Theorem 3.1.5) ;*

g) *for any representation* (Y,u,v) *of* (X,M) *the map* $\hat{u}$ *is an isomorphism of unital algebras ;*

h) $\xi,\eta\in M^\rho \Longrightarrow (|\xi|\wedge|\eta| = 0 \Longleftrightarrow \xi\eta = 0)$;

i) $\eta\in M^\rho$, $\eta^2 = \eta \Longrightarrow \eta\wedge(\dot{1}_X-\eta) = 0$, $\eta\in M_b^\pi$

We prove first the existence. We set

$$\xi\eta = \hat{\varphi}(\xi,\eta)$$

for any $(\xi,\eta)\in M^{\rho}$, where φ is the function defined in f). Let (Y,u,v) be a representation of (X,M) (Theorem 2.3.8). By Proposition 3.1.1 d) we have

$$\hat{u}1'_X = u1_X = 1_Y \quad .$$

Using this result we deduce with the aid of Theorem 3.1.5 that M^{ρ} endowed with the composition law

$$M^{\rho}\times M^{\rho} \longrightarrow M^{\rho} \;,\; (\xi,\eta) \longmapsto \xi\eta$$

is a commutative, associative, unital algebra possessing the properties a), f) and g). b) follows from Theorem 3.1.5 and Proposition 1.5.5 b). c) follows from g) and Proposition 3.1.1 b), c). e) follows from g), Proposition 3.1.1 c) and Proposition 1.6.2. h) and i) follow from g) and Proposition 3.1.1 b).

In order to prove the uniqueness let

$$M^{\rho}\times M^{\rho} \longrightarrow M^{\rho} \;,\; (\xi,\eta) \longmapsto \xi\cdot\eta$$

be a map possessing the above properties 1), 2), 3). We have by 1)

$$\xi\cdot 0 = \xi\cdot(0+0) = \xi\cdot 0 + \xi\cdot 0$$

and therefore

$$\xi\cdot 0 = 0$$

for any $\xi\in M^{\pi}_b$. We deduce by 3)

$$\xi\cdot\eta = \xi\cdot(\eta\vee 0) = (\xi\cdot\eta)\vee(\xi\cdot 0) \geqslant 0$$

for any $\xi,\eta\in (M^{\pi}_b)_+$. By [32] Theorem V 8.2 we get

$$\xi\cdot\eta = \xi\eta$$

for any $\xi,\eta\in M^{\pi}_b$. With the aid of 3) and Corollary 3.1.2 a) we deduce

$$\xi\cdot\eta = \xi\cdot(\bigvee_{n\in\mathbb{N}}(\eta\wedge n1_X)) = \bigvee_{n\in\mathbb{N}}(\xi\cdot(\eta\wedge n1_X)) = \bigvee_{n\in\mathbb{N}}(\xi(\eta\wedge n1_X)) =$$

$$=\xi\left(\bigvee_{n\in\mathbb{N}}(\eta\wedge ni_X)\right) = \xi\eta$$

for any $(\xi,\eta)\in(M_b^\pi)_+\times M_+^\rho$ and further

$$\xi\cdot\eta = \left(\bigvee_{n\in\mathbb{N}}(\xi\wedge ni_X)\right)\cdot\eta = \bigvee_{n\in\mathbb{N}}((\xi\wedge ni_X)\cdot\eta) = \bigvee_{n\in\mathbb{N}}((\xi\wedge ni_X)\eta) =$$

$$= \left(\bigvee_{n\in\mathbb{N}}(\xi\wedge ni_X)\right)\eta = \xi\eta$$

for any $(\xi,\eta)\in M_+^\rho\times M_+^\rho$. Let $(\rho,\xi)\in M^\rho\times M^\rho$. Then

$$\xi\cdot\eta = (\xi_+-\xi_-)\cdot(\eta_+-\eta_-) = \xi_+\cdot\eta_+ - \xi_-\cdot\eta_+ - \xi_+\cdot\eta_- + \xi_-\cdot\eta_- =$$

$$= \xi_+\eta_+ - \xi_-\eta_+ - \xi_+\eta_- + \xi_-\eta_- = \xi\eta. \ \square$$

We shall consider from now on M^ρ, M_c^π, M_b^π, *and* M^π *endowed with the algebraic structures defined in Theorem 3.1.7.*

Remark Via Proposition 3.1.1 and Theorem 2.3.8,e) follows from [16] III Corollary 7.7.

Proposition 3.1.8 *Let* $\xi\in M^\rho$. *We set*

$$F := \{\eta\in M^\rho \mid |\eta| \leq |\xi|\}.$$

Then :

a) $\forall(\eta,\mu)\in F\times M(\xi) \Longrightarrow \eta\in\hat{L}^1(\mu)$;

b) F *endowed with the weakest topology for which the maps*

$$F \longrightarrow \mathbb{R},\ \eta \longmapsto \int\eta\,d\mu \quad (\mu\in M(\xi))$$

are continuous is compact.

a) follows from Corollary 3.1.3 a).

b) By Theorem 2.3.8 the given topology is Hausdorff. Let $\underline{\underline{F}}$ be an ultrafilter on F and let ζ be the map

$$M(\xi) \longrightarrow \mathbb{R}\ ,\ \mu \longmapsto \lim_{\eta,\underline{\underline{F}}} \int \eta d\mu\ .$$

ζ is linear and from

$$\forall \mu \in M(\xi) \Longrightarrow |\zeta(\mu)| \leqslant \int |\xi|\ d\ |\mu|$$

we get $\zeta \in M(\xi)^{\pi} \subset M^{\rho}$. From the above relation we also deduce $\zeta \in F$. Since $\underline{\underline{F}}$ converges to ζ it follows that F is compact. □

2. Spaces associated to a measure

Let $\mu \in M$, let $p \in [1,\infty[$, and let φ_p be the Borel measurable real function on $\bar{\mathbb{R}}$

$$\bar{\mathbb{R}} \longrightarrow \mathbb{R}\ ,\ \alpha \longmapsto \begin{cases} |\alpha|^p & \text{if} \quad \alpha \in \mathbb{R} \\ \infty & \text{if} \quad \alpha \in \bar{\mathbb{R}} \setminus \mathbb{R}\ . \end{cases}$$

For any $\xi \in M^{\rho}$ we set (Theorem 3.1.5)

$$|\xi|^p := \hat{\varphi}_p(\xi)\ .$$

We denote

$$\hat{\mathcal{L}}^p(\mu) := \{\xi \in M^{\rho} \mid |\xi|^p \in \hat{\mathcal{L}}^1(\mu)\}\ ,$$

$$\hat{\mathcal{L}}^p_{loc}(\mu) := \{\xi \in M^{\rho} \mid |\xi|^p \in \hat{\mathcal{L}}^1_{loc}(\mu)\}\ ,$$

$$\hat{\mathrm{L}}^p(\mu) := \hat{\mathcal{L}}^p(\mu) \cap \hat{\mathcal{L}}^{\perp}(\mu)\ ,\ \hat{\mathrm{L}}^p_{loc}(\mu) := \hat{\mathcal{L}}^p_{loc}(\mu) \cap \hat{\mathcal{L}}^{\perp}(\mu)\ .$$

<u>*Proposition 3.2.1*</u> *For any fundamental solid subspace* N *of* M *and for any* $p \in [1,\infty[$ *we have*

$$\bigcap_{\mu \in M} \hat{\mathcal{L}}^p_{loc}(\mu) = \bigcap_{\mu \in M_c} \hat{\mathcal{L}}^p_{loc}(\mu) = \bigcap_{\mu \in M_c} \hat{\mathcal{L}}^p(\mu) = M_c^{\pi}\ ,$$

$$\bigcap_{\mu \in N} \hat{\mathcal{L}}^1(\mu) = \{\xi \in M_c^{\pi} \mid \forall \mu \in N \Longrightarrow \xi \cdot \mu \in M_b\} = N^{\pi}\ .$$

The assertions follow immedaitely from Theorem 2.3.8, Proposition 3.1.1 and Theorem 3.1.5 . □

Theorem 3.2.2 Let $\mu\in M$ and let M_μ be the band generated by μ. Then :

a) $\xi\cdot\mu\in M_\mu$ for any $\xi\in\hat{\mathcal{L}}^1_{loc}(\mu)$;

b) the map

$$\hat{\mathcal{L}}^1_{loc}(\mu) \longrightarrow M_\mu \,, \quad \xi \longmapsto \xi\cdot\mu$$

is surjective, linear, and $1_X\cdot\mu = \mu$; it is a homomorphism of vector lattices if μ is positive ;

c) the map

$$\hat{L}^1_{loc}(\mu) \longrightarrow M_\mu \,, \quad \xi \longmapsto \xi\cdot\mu$$

is bijective; it is an isomorphism of vector lattices if μ is positive;

d) for any $(\xi,\eta)\in M^\rho\times\hat{\mathcal{L}}^1_{loc}(\mu)$ we have

$$\xi\in\hat{\mathcal{L}}^1(\eta\cdot\mu) \Longleftrightarrow \xi\eta\in\hat{\mathcal{L}}^1(\mu) \Longrightarrow \int\xi d(\eta\cdot\mu) = \int\xi\eta d\mu \;;$$

$$\xi\in\hat{\mathcal{L}}^1_{loc}(\eta\cdot\mu) \Longleftrightarrow \xi\eta\in\hat{\mathcal{L}}^1_{loc}(\mu) \Longrightarrow \xi\cdot(\eta\cdot\mu) = (\xi\eta)\cdot\mu \,,$$

$$\eta\in\hat{\mathcal{L}}^1(\mu) \Longleftrightarrow \eta\cdot\mu\in M_b \Longrightarrow \int\eta d\mu = \int 1_X d(\eta\cdot\mu) \;;$$

e) there exists $\xi,\eta\in(M_b^\pi)_+\cap\hat{L}^1_{loc}(\mu)$ such that

$$\xi^2 = \xi \,, \; \eta^2 = \eta \,, \; \xi\eta = \xi\wedge\eta = 0 \,,$$

$$\xi\cdot\mu = \mu_+,\; \eta\cdot\mu = -\mu_-,\; (\xi-\mu)\cdot\mu = |\mu| \;;$$

f) $\hat{\mathcal{L}}^1_{loc}(\mu) = \hat{\mathcal{L}}^1_{loc}(|\mu|)$ and $|\xi\cdot\mu| = |\xi|\cdot|\mu|$ for any $\xi\in\hat{\mathcal{L}}^1_{loc}(\mu)$;

g) $\xi\eta\in\hat{\mathcal{L}}^1(\mu)$ for any $(\xi,\eta)\in\hat{\mathcal{L}}^1_{loc}(\mu)\times M^\pi$.

The assertions follow from Theorem 2.3.8, Proposition 3.1.1, and Theorem 3.1.7. □

Proposition 3.2.3 Let $\mu\in M$ and let φ be a real function on $\overline{\mathbb{R}}$ such that the map

$$\mathbb{R} \longrightarrow \mathbb{R}\ ,\ \alpha \longrightarrow \varphi(\alpha)$$

is increasing and bijective. Then φ *is Borel measurable and for any* $\xi \in L^{\perp}(\mu)$ *for which* $\hat{\varphi}(\xi) \in \hat{L}^{1}(\mu)$ *there exists a measurable real function* f *on* X *such that :*

a) $\varphi \circ f \in L^{1}(\mu)$;

b) ξ *is the component of* $\dot{f}$ *on* $\hat{L}^{\perp}(\mu)$.

It is obvious that φ is Borel measurable. By Theorem 3.2.2 d) $\hat{\varphi}(\xi)\cdot\mu \in M_b$. By Proposition 1.4.1 $\hat{\varphi}(\xi)\cdot\mu$ possesses a countable concassage and therefore (Proposition 1.4.4 and Theorem 3.2.2 a)) there exists a measurable real function $g \in L^{1}_{loc}(\mu)$ with $g\cdot\mu = \hat{\varphi}(\xi)\cdot\mu$. Since $g\cdot\mu \in M_b$ we have $g \in L^{1}(\mu)$. For any x let f(x) be the real number for which $\varphi(f(x)) = g(x)$. It is obvious that f is a measurable real function on X such that $\varphi \circ f = g \in L^{1}(\mu)$. By Theorem 3.1.5

$$\dot{g} = \widehat{\varphi \circ f}^{\,\cdot} = \hat{\varphi}(\dot{f}) \ .$$

By Theorem 3.1.5 $\hat{\varphi}(\xi) \in \hat{L}^{\perp}(\mu)$ and from $\hat{\varphi}(\xi)\cdot\mu = \hat{\varphi}(\dot{f})\cdot\mu$ we deduce that $\hat{\varphi}(\xi)$ is the component of $\hat{\varphi}(\dot{f})$ on $\hat{L}^{\perp}(\mu)$. Again by Theorem 3.1.5 we see that ξ is the component of $\dot{f}$ on $\hat{L}^{\perp}(\mu)$. □

<u>*Theorem 3.2.4*</u> *Let* $\mu \in M$ *and let* $p \in [1,\infty[$. *Then :*

a) $\hat{L}^{p}(\mu)$ *and* $\hat{L}^{p}_{loc}(\mu)$ *are fundamental solid subspaces of* M^{ρ} ;

b) $\hat{L}^{p}(\mu)$ *is a fundamental solid subspace of* $\hat{L}^{\perp}(\mu)$;

c) *the map*

$$\hat{L}^{p}(\mu) \longrightarrow \mathbb{R}\ ,\ \xi \longmapsto \left(\int |\xi|^{p} d\mu\right)^{1/p}$$

is a lattice seminorm (an L-seminorm if p = 1) ;

d) *the map*

$$\hat{L}^{p}(\mu) \longrightarrow \mathbb{R}\ ,\ \xi \longmapsto \left(\int |\xi|^{p} d\mu\right)^{1/p}$$

is a lattice norm and $\hat{L}^{p}(\mu)$ *endowed with it is an order complete Banach lattice ;*

e) $f \in L^p(\mu) \Longleftrightarrow \dot{f} \in \hat{L}^p(\mu)$ *for any* $f \in L_\infty$;

f) if for any $f \in L^p(\mu) \cap L_\infty$ *we denote by* $\tilde{f}$ *the component of* $\dot{f}$ *on* $\hat{L}^\perp(\mu)$ *then the map*

$$L^p(\mu) \cap L_\infty \longrightarrow \hat{L}^p(\mu) , \ f \longmapsto \tilde{f}$$

is surjective and linear and we have

$$f \geq 0 \ \mu\text{- a.e.} \Longleftrightarrow \tilde{f} \geq 0 ,$$

$$\int |f|^p d\mu = \int |\tilde{f}|^p d\mu$$

for any $f \in L^p(\mu) \cap L_\infty$ *; the map* $L^p(\mu) \longrightarrow \hat{L}^p(\mu)$ *defined by it is an isomorphism of normed vector lattices.*

Let (Y,u,v) be a representation of (X,M) (Theorem 2.3.8). By Proposition 3.1.1 a) and Theorem 3.1.5 we get all assertions up to the surjectivity of the map

$$L^p(\mu) \cap L_\infty \longrightarrow \hat{L}^p(\mu) , \ f \longmapsto \tilde{f}$$

which follows from Proposition 3.2.3. □

<u>*Proposition 3.2.5*</u> *The map*

$$\tilde{\mu} : \hat{L}^1(\mu) \longrightarrow \mathbb{R} , \ \xi \longmapsto \int \xi d\mu$$

belongs to $(\hat{L}^1(\mu))^\pi$ *and therefore to* $M^{\rho\rho}$ *for any* $\mu \in M$. *The map*

$$M \longrightarrow M^{\rho\rho} , \ \mu \longmapsto \tilde{\mu}$$

is injective, linear, we have

$$\mu \geq 0 \Longleftrightarrow \tilde{\mu} \geq 0$$

for any $\mu \in M$, *and* $\{\tilde{\mu} | \mu \in M\}$ *is a fundamental solid subspace of* $M^{\rho\rho}$.

By Theorem 3.2.4 a) $\hat{L}^1(\mu)$ is a fundamental solid subspace of M^ρ and therefore $(\hat{L}^1(\mu))^\pi \subset M^{\rho\rho}$ for any $\mu \in M$. It is obvious that

$\mu\in(\hat{\mathcal{L}}^1(\mu))^\pi$ for any $\mu\in M$ and that the map

$$M \longrightarrow M^{\rho\rho}, \ \mu \longmapsto \tilde{\mu}$$

is injective, linear, and that $\mu \geqslant 0$ iff $\tilde{\mu} \geqslant 0$. Let $\zeta\in M^{\rho\rho}$ such that there exists $\mu\in M$ with $|\zeta|\leqslant|\tilde{\mu}|$. Let (Y,u,v) be a representation of (X,M) (Theorem 2.3.8). Then ζ is defined on the set $K(Y)$ of continuous real functions on Y with compact carrier and therefore there exists $\nu\in M$ such that

$$\langle f,\zeta\rangle = \int f d(v\nu)$$

for any $f\in K(Y)$. It is easy to see that the above relation implies $\zeta = \tilde{\nu}$. Hence $\{\tilde{\mu}\,|\,\mu\in M\}$ is a solid subspace of $M^{\rho\rho}$ and it is not difficult to see that it is fundamental. □

<u>Remark</u> The evaluation map $M^\rho \longrightarrow M^{\rho\rho\rho}$ is an isomorphism of vector lattices.

<u>*Proposition 3.2.6*</u> *Let* $\mu,\nu\in M$, *let* $\alpha,\beta\in\mathbb{R}$, *and let* $p\in[1,\infty[$. *Then:*

a) $\hat{\mathcal{L}}^p(\mu)\cap\hat{\mathcal{L}}^p(\nu)\subset\hat{\mathcal{L}}^p(\alpha\mu+\beta\nu)$ *and*

$$\int\xi d(\alpha\mu+\beta\nu) = \alpha\int\xi d\mu+\beta\int\xi d\nu$$

for any $\xi\in\hat{\mathcal{L}}^1(\mu)\cap\hat{\mathcal{L}}^1(\nu)$;

b) $\hat{\mathcal{L}}^p_{loc}(\mu)\cap\hat{\mathcal{L}}^p_{loc}(\nu)\subset\hat{\mathcal{L}}^p_{loc}(\alpha\mu+\beta\nu)$ *and*

$$\xi\cdot(\alpha\mu+\beta\nu) = \alpha(\xi\cdot\mu) + \beta(\xi\cdot\nu)$$

for any $\xi\in\hat{\mathcal{L}}^1_{loc}(\mu)\cap\hat{\mathcal{L}}^1_{loc}(\nu)$;

c) $|\mu|\leqslant|\nu| \Longrightarrow \hat{\mathcal{L}}^p(\nu)\subset\hat{\mathcal{L}}^p(\mu)$, $\hat{\mathcal{L}}^p_{loc}(\nu)\subset\hat{\mathcal{L}}^p_{loc}(\mu)$;

d) for any $\mu,\nu\in M$ *with* $|\mu|\wedge|\nu| = 0$ *we have*

$$\hat{\mathcal{L}}^p(\mu+\nu) = \hat{\mathcal{L}}^p(\mu)\cap\hat{\mathcal{L}}^p(\nu)\ ,\ \hat{\mathcal{L}}^p_{loc}(\mu+\nu) = \hat{\mathcal{L}}^p_{loc}(\mu)\cap\hat{\mathcal{L}}^p_{loc}(\nu)\ ,$$

$$\hat{L}^p(\mu+\nu) = \hat{L}^p(\mu) + \hat{L}^p(\nu)\ ,\ \hat{L}^p_{loc}(\mu+\nu) = \hat{L}^p_{loc}(\mu) + \hat{L}^p_{loc}(\nu)\ .$$

All assertions follow immediately from Theorem 2.3.8, Proposition

3.1.1 a), b), e), and Theorem 3.1.5. □

Proposition 3.2.7 *Let* $(\mu_\iota)_{\iota\in I}$ *be an upper bounded family in* M *and let* ξ *be a positive element of*

$$\hat{\mathcal{L}}^1_{\text{loc}}(\bigvee_{\iota\in I}\mu_\iota)\cap(\bigcap_{\iota\in I}\hat{\mathcal{L}}^1_{\text{loc}}(\mu_\iota)) .$$

We have :

a) $\xi\cdot(\bigvee_{\iota\in I}\mu_\iota) = \bigvee_{\iota\in I}(\xi\cdot\mu_\iota)$.

b) *if* $(\mu_\iota)_{\iota\in I}$ *is upper directed and*

$$\xi\in\hat{\mathcal{L}}^1(\bigvee_{\iota\in I}\mu_\iota)\cap(\bigcap_{\iota\in I}\hat{\mathcal{L}}^1(\mu_\iota))$$

then

$$\int\xi d(\bigvee_{\iota\in I}\mu_\iota) = \sup_{\iota\in I}\int\xi d\mu_\iota .$$

The assertions follow from Theorem 2.3.8 and Proposition 3.1.1. □

Proposition 3.2.8 *Let* N *be a countable subset of* M *and let*

$$\xi\in\bigcap_{\mu\in N}\hat{\mathcal{L}}^1_{\text{loc}}(\mu) .$$

If there exists a locally countable, disjoint family $(A_\iota)_{\iota\in I}$ *in* $\underline{\underline{R}}$ *such that*

$$|\xi| = \sum_{\iota\in I}|\xi|\dot{1}_{A_\iota}$$

then there exists a measurable real function f *on* X *such that* $\dot{f}\cdot\mu = \xi\cdot\mu$ *for any* $\mu\in N$ *and* $f = f1_A$ *, where* $A := \bigcup_{\iota\in I}A_\iota$.

Let (Y,u,v) be a representation of (X,M) (Theorem 2.3.8). By Proposition 3.1.1 e) $\hat{u}\xi\in\bigcap_{\mu\in N}L^1_{\text{loc}}(v\mu)$. By the hypothesis $(A_\iota)_{\iota\in I}$ is a locally countable concassage of $\xi\cdot\mu$ for any $\mu\in N$ and therefore by Proposition 2.4.4 d) there exists a measurabel real function g on X

such that $ug = \hat{u}\xi$ on $\bigcup_{\mu\in N} \mathrm{Supp}(v\mu)$. We set $f = g1_A$. Then f is a measurable real function on X such that

$$\dot{f}\cdot\mu = (\dot{g}\dot{1}_A)\cdot\mu = (\xi\dot{1}_A)\cdot\mu = \xi\cdot\mu$$

for any $\mu\in N$. □

<u>*Proposition 3.2.9*</u> *Let $(\mu_\iota)_{\iota\in I}$ be a family in M and let $\mu\in M$ such that*

$$\{\xi\in\bigcap_{\iota\in I}\hat{L}^1(\mu_\iota) \mid \xi^2 = \xi\}\subset\hat{L}^1(\mu) .$$

If $|\mu_\iota|\wedge|\mu| = 0$ for any $\iota\in I$ then $\mu\in M_b$.

Let (Y,u,v) be a representation of (X,M) (Theorem 2.3.8) and let F be the closure of $\bigcup_{\iota\in I} \mathrm{Supp}(v\mu_\iota)$. Then

$$F\cap \mathrm{Supp}(v\mu) = \emptyset .$$

Since

$$\hat{u}^{-1}1^Y_{Y\setminus F}\in\bigcap_{\iota\in I}\hat{L}^1(\mu_\iota) \quad , \quad \hat{u}^{-1}1^Y_F\in\hat{L}(\mu)$$

we get $1_X\in L^1(\mu)$ (Proposition 3.1.1 a), d)). Hence $\mu\in M_b$. □

3. The spaces M, M_b, and M_c

<u>*Theorem 3.3.1*</u> *The evaluation maps*

$$M \longrightarrow M^{\pi\pi} , \quad M_b \longmapsto M_b^{\pi\pi}$$

are isomorphisms of vector lattices. The evaluation map

$$M_c \longrightarrow M_c^{\pi\pi}$$

is injective and its image is a fundamental solid subspace of $M_c^{\pi\pi}$.

Let (Y,u,v) be a representation of (X,M) (Theorem 2.3.8). By

Definition 2.3.1 e) and Proposition 2.3.2 b) M, M_b, and M_c are isomorphic as vector lattices to $M(Y)$, $M_b(Y)$, and $M_c(Y)$ respectively. The assertions now follow from Proposition 1.6.1 c), d). □

Remark. By the remark of Proposition 1.6.1 the evaluation map $M_c \longrightarrow M_c^{\pi\pi}$ is not always bijective.

Theorem 3.3.2 *We have* $M_b^{\pi} = M_b^{+}$, $M_c^{\pi} = M_c^{+}$.

The first assertions follows e.g. from [26] Theorem V 8.6. Let (Y,u,v) be a representation of (X,M) (Theorem 2.3.8). By Definition 2.3.1 e) and Proposition 2.3.2 b) M_c is isomorphic as vector lattice to $M_c(Y)$. Hence $M_c^{\pi} = M_c^{+}$. □

Remark. M^{π} and M^{+} may be different as the following example shows. We take $X := \mathbb{N}$, $\underline{\underline{R}} := \{A \subset \mathbb{N} \mid \sum_{n\in A} \frac{1}{n} < \infty\}$, and M the set of all measures on $\underline{\underline{R}}$. Then for any ultrafilter $\underline{\underline{F}}$ on $\mathbb{N}$ with $\underline{\underline{R}} \cap \underline{\underline{F}} = \emptyset$ the map

$$M \longrightarrow \mathbb{R}, \quad \mu \longmapsto \lim_{n,\underline{\underline{F}}} n\mu(\{n\})$$

belongs to $M^{+} \setminus M^{\pi}$.

Proposition 3.3.3 *If* $M = M_b$ *then* $M_b^{\pi} = M_c^{\pi}$.

Let (Y,u,v) be a representation of (X,M) (Theorem 2.3.8). By Definition 2.3.1 e) and Proposition 2.3.2 b) M, M_b, and M_c are canonically isomorphic to $M(Y)$, $M_b(Y)$, and $M_c(Y)$ respectively. By Proposition 1.6.3 $C(Y) = C_b(Y)$. By Proposition 1.6.1 a), b) $C(Y)$ and $C_b(Y)$ are canonically isomorphic to $M_c(Y)^{\pi}$ and $M_b(Y)^{\pi}$ respectively. Hence $M_c^{\pi} = M_b^{\pi}$. □

Remark. The example given in the preceding remark shows that neither $M = M_b$ nor $M_b = M_c$ follow from $M_b^{\pi} = M_c^{\pi}$.

Proposition 3.3.4 *If there exists* $\xi \in M^{\pi}$ *such that for any* $\eta \in M^{\pi}$ *there exists* $\alpha \in \mathbb{R}_+$ *with* $|\eta| \leq \alpha|\xi|$ *then* $M = M_b$ *and* $M_b^{\pi} = M_c^{\pi}$.

Let (Y,u,v) be a representation of (X,M) (Theorem 2.3.8). By

Proposition 3.1.1 b) c) $\hat{u}\xi$ does not vanish on Y and therefore

$$\mathrm{Supp}\ (\hat{u}\xi) = Y .$$

By Proposition 1.6.2 $1_Y \in C_i(Y)$ and therefore (Proposition 3.1.1 c), d)) $1_X \in M^\pi$. Hence $M = M_b$. By Proposition 3.3.3 $M_b^\pi = M_c^\pi$. □

Proposition 3.3.5 *For any $\varphi \in M^{\rho +}$ there exists a finite family $(\varphi_\iota)_{\iota \in I}$ of positive ring homomorphismus $M^\rho \longrightarrow \mathbb{R}$ and a family $(\alpha_\iota)_{\iota \in I}$ of real numbers such that*

$$\varphi = \sum_{\iota \in I} \alpha_\iota \varphi_\iota$$

Let (Y,u,v) be a representation of (X,M) (Theorem 2.3.8). By Proposition 3.1.1 b) and Theorem 3.1.7 g) M^ρ is isomorphic to $C_\infty(Y)$ as vector lattice and as algebra and the assertion follows from Proposition 1.6.6. □

Proposition 3.3.6 *Let $\varphi \in M^{\pi\pi}$ and let (Y,u,v) be a representation of (X,M). The following assertions are equivalent :*

a) *φ is a ring homomorphism ;*
b) *φ is a lattice homomorphism and $\varphi \dot{1}_X = 1$;*
c) *φ belongs to an extremal ray of $(M^{\pi\pi})_+$ and $\varphi \dot{1}_X = 1$;*
d) *there exists an isolated point y of Y such that*

$$\varphi\xi = (\hat{u}\xi)(y)$$

for any $\xi \in M^\pi$;

e) *φ is the restriction to M^π of an element of $M^{\rho\pi}$ which is a ring and a lattice homomorphism and which belongs to an extremal ray of $(M^{\rho\pi})_+$.*

a $\Rightarrow$ d, b $\Rightarrow$ d, c $\Rightarrow$ d. By Theorem 3.3.1 there exists $\mu \in M$ such that

$$\varphi\xi = \int \xi d\mu$$

for any $\xi \in M^\pi$. By Proposition 3.1.1 a) we get further

$$\varphi\xi = \int \hat{u}\xi d(\nu\mu)$$

for any $\xi \in M^{\pi}$ and this implies d) .

d $\Longrightarrow$ e and e $\Longrightarrow$ a & b & c is obvious. □

Proposition 3.3.7 Let $\hat{X}$ be the set of elements of $M^{\pi\pi}$ which are ring homomorphismus. For any $A \in \underline{\underline{R}}$ we set

$$\hat{A} := \{\hat{x} \in \hat{X} \mid x(\dot{1}_A) = 1\}$$

and set $\hat{\underline{\underline{R}}} := \{\hat{A} \mid A \in \underline{\underline{R}}\}$, Then :

a) $A, B \in \underline{\underline{R}} \Longrightarrow \widehat{A \cup B} = \hat{A} \cup \hat{B}, \quad \widehat{A \setminus B} = \hat{A} \setminus \hat{B}$;

b) $(A_n)_{n \in \mathbb{N}}$ sequence in $\underline{\underline{R}} \Longrightarrow \widehat{\bigcap_{n N} A_n} = \bigcap_{n \in \mathbb{N}} \hat{A}_n$;

c) $\hat{\underline{\underline{R}}}$ is a δ-ring .

a) and b) follow immediately from Theorem 2.3.8 and Proposition 3.3.6 a $\Longleftrightarrow$ d. c) follows from a & b. □

Proposition 3.3.8 With the notations of Proposition 3.3.7 we set

$$\hat{M}(\underline{\underline{R}}) := \{\mu \in M(\underline{\underline{R}}) \mid A \in \underline{\underline{R}}, \ \hat{A} = \emptyset \Longrightarrow \mu(A) = 0\}$$

and denote for any $\lambda \in M(\hat{\underline{\underline{R}}})$ by $\hat{\lambda}$ the map

$$\underline{\underline{R}} \longrightarrow \mathbb{R}, \ A \longmapsto \lambda(\hat{A}) .$$

Then :

a) $\hat{M}(\underline{\underline{R}})$ is a band of $M(\underline{\underline{R}})$;

b) $\hat{\lambda} \in \hat{M}(\underline{\underline{R}})$ for any $\lambda \in M(\hat{\underline{\underline{R}}})$;

c) the map

$$M(\hat{\underline{\underline{R}}}) \longrightarrow \hat{M}(\underline{\underline{R}}), \ \lambda \longmapsto \hat{\lambda}$$

is an isomorphism of vector lattices.

a) is obvious.

b) follows from Proposition 3.3.7 a), b).

c) It is obvious that λ is positive iff $\hat{\lambda}$ is positive and that the map

$$M(\hat{\underline{\underline{R}}}) \longrightarrow \hat{M}(\underline{\underline{R}}) \ , \ \lambda \longmapsto \hat{\lambda}$$

is linear and injective. Let $\mu \in \hat{M}(\underline{\underline{R}})$, and let $A, B \in \underline{\underline{R}}$ such that $\hat{A} = \hat{B}$. By Proposition 3.3.7 a) we have

$$\widehat{A \setminus B} = \widehat{B \setminus A} = \emptyset$$

and therefore $\mu(A) = \mu(B)$. Hence there exists a real function λ on $\hat{\underline{\underline{R}}}$ such that $\lambda(\hat{A}) = \mu(A)$ for any $A \in \underline{\underline{R}}$. Let $(B_n)_{n \in \mathbb{N}}$ be a disjoint sequence in $\hat{\underline{\underline{R}}}$ whose union B belongs to $\hat{\underline{\underline{R}}}$. Let $(A_n)_{n \in \mathbb{N}}$ be a sequence in $\underline{\underline{R}}$ such that $\hat{A}_n = B_n$ for any $n \in \mathbb{N}$ and let $A \in \underline{\underline{R}}$ with $\hat{A} = B$. For any $n \in \mathbb{N}$ we set

$$A'_n := A \cap (A_n \setminus \bigcup_{m < n} A_m) .$$

Then $(A'_n)_{n \in \mathbb{N}}$ is a disjoint sequence in $\underline{\underline{R}}$ whose union belongs to $\underline{\underline{R}}$ and by Proposition 3.3.7 a), b) $\hat{A}'_n = B_n$ for any $n \in \mathbb{N}$ and

$$\widehat{\bigcup_{n \in \mathbb{N}} A'_n} = B .$$

Hence $\lambda \in M(\hat{\underline{\underline{R}}})$ and $\hat{\lambda} = \mu$. □

Proposition 3.3.9 *Let* F *be a solid subspace of* M_c^π *containing* $\{i_A \mid A \in \underline{\underline{R}}\}$ *and let* N *be the set* $\bigcap_{\xi \in F} M(\xi)$. *Then* N *is a fundamental solid subspace of* M, $F \subset N^\pi$, *and the evaluation map* $N \longrightarrow F^{\pi\pi}$ *and the map*

$$N^{\pi\pi} \longrightarrow F^\pi \ , \ \varphi \longmapsto \varphi | F^\pi$$

are isomprphismus of vector lattices.

It is obvious that N is a fundamental solid subspace of M and that $F \subset N^\pi$. Let (Y,u,v) be a representation of (X,M) (Theorem 2.3.8) and let $\psi \in F^\pi$. By Proposition 3.1.1 a), b), c) there exists $\mu \in M$ such that

$$\int f d(v\mu) = \psi(\hat{u}^{-1}f)$$

for any $f \in C_c(Y)$. We deduce (Proposition 3.1.1 a) b) c)) $\xi \in \hat{L}^1(\mu)$ and

$$\xi(\mu) = \int(\hat{u}\xi)\, d(v\mu) = \psi(\xi)$$

for any $\xi \in F$ = Hence $\mu \in N$ and the evaluation map $N \longrightarrow F^{\pi\pi}$ is surjective. It follows immediately that the map

$$N^{\pi\pi} \longrightarrow F^{\pi}, \ \varphi \longrightarrow \varphi | F^{\pi}$$

is surjective too. We deduce now easily that both maps are isomorphismic of vector lattices. □

4. Structures on M and M_c^{π}

Definition 3.4.1 *For any fundamental solid subspace* N *of* M *and for any* $\xi \in N^{\pi}$ *we denote by* q_{ξ} *the* L-*seminorm on* N

$$N \longrightarrow \mathbb{R}, \ \mu \longmapsto \int |\xi| d|\mu|$$

(Proposition 3.2.1); *for any subset* F *of* N^{π} *we denote by* (N,F) *the space* N *endowed with the topology generated by the set* $\{q_{\xi} | \xi \in F\}$ *of seminorms* (if F generates N^{π} as band then (N,F) is Hausdorff). *Let* G *be a solid subspace of* M_c^{π} ; *we denote by* $\tilde{G}$ *the set of* $\xi \in M^{\rho}$ *such that for any* $\eta \in G$ *there exists* $\alpha \in R_+$ *with* $|\xi\eta| \leqslant \alpha|\xi|$. *For any* $\xi \in \tilde{G}$ *we denote by* p_{ξ} *the* M-*seminorm on* G

$$G \longrightarrow \mathbb{R}, \ \eta \longmapsto \inf\{\alpha \in R_+ | \ |\xi\eta| \leqslant \alpha|\xi|\};$$

for any subset F *of* $\tilde{G}$ *we denote by* (G,F) *the space* G *endowed with the topology generated by the set* $\{p_{\xi} | \xi \in F\}$ *of seminorms* ($\tilde{G}$ is a solid subpsace of M^{ρ}; if F generates $\tilde{G}$ as band then (G,F) is Hausdorff).

By Theorem 2.3.8 and Proposition 1.6.5 and 3.1.1 e) $M^{\pi} \subset \tilde{M}_c^{\pi}$.

Proposition 3.4.2 *Let* N *be a fundamental solid subspace of* M, *let* F *be a subset of* N^{π}, *and let* G *be the solid subspace of* N^{π}

generated by F. *Then* :

a) the topology of (N,F) *is the topology of uniform convergence on the order bounded sets of* G ;

b) G *is the dual of* (N,F) ;

c) if the evaluation map $N \longrightarrow G^{\pi}$ *is bijective then* (N,F) *is topologically complete.*

a) We set

$$U_{\xi} := \{\mu\in N \mid \int|\xi|d|\mu| \leqslant 1\}$$

$$V_{\eta} := \{\mu\in N \mid \zeta\in G, |\zeta| \leqslant |\eta| \Longrightarrow |\int\zeta d\mu| \leqslant 1\}$$

for any $\xi\in F$, $\eta\in G$. Since $U_{\xi} = V_{\xi}$ for any $\xi\in F$ it follows that the topology on N of uniform convergence on the order bounded sets of G is finer than the topology of (N,F) . Let $\eta\in G$. There exists a finite family $(\xi_{\iota},\alpha_{\iota})_{\iota\in I}$ in $F\times\mathbb{R}_{+}$ such that

$$|\eta| \leqslant \sum_{\iota\in I} \alpha_{\iota}|\xi_{\iota}| .$$

We get

$$|\int\zeta d\mu| \leqslant \sum_{\iota\in I} \alpha_{\iota}\int|\xi_{\iota}|d|\mu| = \sum_{\iota\in I} \alpha_{\iota}q_{\xi_{\iota}}(\mu)$$

for any $\zeta\in G$ with $|\zeta| \leqslant |\eta|$ and therefore

$$\frac{1}{1+\sum_{\iota\in I}\alpha_{\iota}} \bigcap_{\iota\in I} U_{\xi_{\iota}} \subset V_{\eta} .$$

Hence the topology of (N,F) is finer than the topology of uniform convergence on the order bounded sets of G .

b) The order bounded sets of G are relatively compact for the topology of pointwise convergence and therefore by a) and Mackey's theorem G is the dual of (M,F_{+}) .

c) Let E be a vector lattice. Then E^{+} endowed with the topology of uniform convergence on the order bounded sets of E is complete. Since E^{π} is a band of E^{+} , E^{π} is closed and therefore complete with respect to the above topology. If we identify N with G^{π} via the evaluation map then by a) and by the above considerations (N,F) is complete. □

Theorem 3.4.3 *If* F *is a solid subspace of* M^{π} *containing* $\{\dot{1}_A | A \in \underline{\underline{R}}\}$ *then :*

a) (M,F) *is a topologically complete Hausdorff locally convex lattice whose dual is* F ;

b) M_c *and* M_b *are dense subspace of* (M,F) ;

c) *if* $(\mu_\iota)_{\iota\in I}$ *is an upper directed family in* M_+ *such that*

$$\sup_{\iota\in I} q_\xi(\mu_\iota) < \infty$$

for any $\xi\in F$ *then* $\bigvee_{\iota\in I} \mu_\iota$ *exists ;*

d) *if* (M,M^{π}) *is normable then* $M = M_b$ *and* $M_b^{\pi} = M_c^{\pi}$.

a) By Proposition 3.3.9 the evaluation map $M \longrightarrow F^{\pi}$ is an isomorphism of vector lattices and the assertion follows from Proposition 3.4.2.

b) Let us order $\underline{\underline{R}}$ by the inclusion relation and let $\underline{\underline{F}}$ be its section filter. Then for any $(\xi,\mu)\in F\times M$ we have

$$\lim_{A,\underline{\underline{F}}} \int \xi d(1_A \cdot \mu) = \int \xi d\mu$$

and therefore

$$\lim_{A,\underline{\underline{F}}} q_\xi(\mu - 1_A \cdot \mu) = 0 .$$

Hence

$$\lim_{A,\underline{\underline{F}}} 1_A \cdot \mu = \mu$$

and therefore M_c and M_b are dense subspaces of (M,F) .

c) let us denote by φ the map

$$F_+ \longrightarrow \mathbb{R}\ ,\ \xi \longrightarrow \sup_{\iota\in I} \xi(\mu_\iota)\ .$$

Then there exists $\psi\in F^\pi$ whose restriction to F_+ equals φ. By Proposition 3.3.9 that we deduce that $\bigvee_{\iota\in I} \mu_\iota$ exists.

d) Assume (M,M^π) is normable. Then there exists $\xi\in M^\pi$ such that for any $\eta\in M^\pi$ there exists $\alpha\in\mathbb{R}_+$ with $|\eta| \leqslant \alpha|\xi|$. By Proposition 3.3.4 $M = M_b$ and $M_c^\pi = M_b^\pi$. □

<u>Remarks</u>. 1. We cannot replace in d) the hypothesis "(M,M^π) is normable" with the hypothesis "(M,M^π) is metrizable" or "there exists $\xi\in M^\pi$ such that q_ξ is a norm". A counterexample for the first conjecture is given by $X := \mathbb{N}$, $\underline{\underline{R}} := \{A|A \text{ finite, } A\subset\mathbb{N}\}$, M = {the set of measures on $\underline{\underline{R}}$}, and for the second by $X := \mathbb{N}$,

$$\underline{\underline{R}} := \{A\subset\mathbb{N} \mid \sum_{n\in A} \frac{1}{n} < \infty\}\ ,$$

M = {the set of measures on $\underline{\underline{R}}$} .

2. Let Σ be the set of σ-rings $\underline{\underline{S}} \subset \underline{\underline{R}}$. By Proposition 1.5.5 d) $\bigvee_{A\in\underline{\underline{S}}} \dot{1}_A$ exists in M^π for any $\underline{\underline{S}}\in\Sigma$; we denote it by $\xi_{\underline{\underline{S}}}$. Let F', F'' be the solid subspaces of M^π generated by $\{\xi_{\underline{\underline{S}}} | \underline{\underline{S}}\in\Sigma\}$ and $\{\dot{1}_A | A\in\underline{\underline{R}}\}$ respectively. The topology of (M,F') is nothing else but the seminorm topology introduced in [5] page 144. The following example will show that the topology of (M,M^π) may be strictly finer than the topology of (M,F'). We take as X a set of cardinal number $\aleph_{\omega_0}$, set

$$\underline{\underline{R}} := \{A\subset X \mid \text{ cardinal of } A < \aleph_{\omega_0}\}\ ,$$

and denote by M the set of measures on $\underline{\underline{R}}$. Then any $\mu\in M$ is bounded and therefore $\dot{1}_X\in M^\pi$ but the seminorm $q_{\dot{1}_X}$ is not continuous on (M,F'). The following example will show that the topology of (M,F') may be strictly finer then the topology of (M,F''). We take as X an uncountable set, as $\underline{\underline{R}}$ the set of countable subsets of X, and as M the set of measures on $\underline{\underline{R}}$. Then $\underline{\underline{R}}$ is a σ-ring and therefore (M,F') is normable while (M,F'') is not.

3. From the example given in the remark of Theorem 3.3.2 it

can be seen that (M,M^{π}) is not always barreled. Indeed let us denote for any $n\in\mathbb{N}$ by f_n the map

$$\mathbb{N} \longrightarrow \mathbb{R}\ ,\ m \longrightarrow \begin{cases} 1 & \text{if } 2^n \leq m < 2^{n+1} \\ 0 & \text{otherwise} \end{cases} .$$

We set $F := \{\dot{f}_n \mid n\in\mathbb{N}\}$ and denote by λ the element of M defined by

$$\underline{\underline{R}} \longrightarrow \mathbb{R}\ ,\ A \longmapsto \sum_{n\in A} \frac{1}{n} .$$

Then $F \subset M^{\pi}$. We want to show that the absolute polar F° of F in M is a barrel of M. The only non-trivial property to be checked is that F° is radial. Let $\mu\in M$. There exists $\nu\in M_b$ and $\alpha\in\mathbb{R}_+$ such that

$$|\mu-\nu| \leq \alpha\lambda$$

and therefore

$$|\mu| \leq |\nu| + \alpha\lambda .$$

We have

$$\dot{f}_n(\lambda) = \sum_{m=2^n}^{2^{n+1}-1} \frac{1}{m} \in \left]\frac{1}{2},1\right[$$

for any $n\in\mathbb{N}$ and therefore

$$\sup_{n\in\mathbb{N}} \dot{f}_n(|\mu|) \leq \|\nu\| + \alpha .$$

Hence F° is radial and therefore a barrel.

We want to show that F° is not a 0-neighbourhood in (M,M^{π}). Assume the contrary. Then there exists an $f\in\mathbb{R}_+^{\mathbb{N}}$ such that $\dot{f}\in M^{\pi}$ and

$$\{\mu\in M \mid \dot{f}(|\mu|) \leq 1\} \subset F^{\circ}.$$

For any $n\in\mathbb{N}$ we set

$$\lambda_n : \underline{\underline{R}} \longrightarrow \mathbb{R}\ ,\ A \longmapsto \sum_{\substack{m\in A_n \\ m\geq 2}} \frac{1}{m} .$$

Then

$$\lim_{n\to\infty} \dot{f}(\lambda_n) = 0 .$$

From

$$\dot{f}\left(\frac{\lambda_n}{\dot{f}(\lambda_n)+\frac{1}{3}}\right) < 1$$

we get

$$\frac{\lambda_n}{\dot{f}(\lambda_n)+\frac{1}{3}} \in F^0$$

and therefore

$$\frac{1}{2} < \dot{f}_n(\lambda_n) \leqslant \dot{f}(\lambda_n) + \frac{1}{3}$$

and this leads to the contradictory relation $\frac{1}{2} \leqslant \frac{1}{3}$.

Proposition 3.4.4 *Let N be a fundamental solid subspace of M, let $(\mu_\iota)_{\iota\in I}$ be a lower directed nonempty family in N, let $\underline{\underline{F}}$ be its lower section filter on N, and let F be a subset of N^π generating N^π as band. Then the following assertions are equivalent:*

a) $\bigwedge_{\iota\in I} \mu_\iota = 0$;

b) $\underline{\underline{F}}$ *converges to 0 in* (N,F).

a $\Longrightarrow$ b follows from Proposition 3.2.1 and 3.2.7.

b $\Longrightarrow$ a. Let ξ belong to the solid subspace of N^π generated by F . Then q_ξ is continuous on (N,F) and therefore

$$0 = \lim_{\mu,\underline{\underline{F}}} (-\int|\xi|d|\mu|) \leqslant \inf_{\iota\in I} \int|\xi|d\mu_\iota \leqslant \lim_{\mu,\underline{\underline{F}}} \int|\xi|d\mu = 0 .$$

We get

$$\inf_{\iota\in I} \int|\xi|d_{\mu_\iota} = 0$$

and therefore

$$\int|\xi|d\mu_\iota \geq 0$$

for any $\iota \in I$. Since F generates N^π as band we deduce $\mu_\iota \geq 0$ for any $\iota \in I$. By a $\Rightarrow$ b $\underline{\underline{F}}$ converges to $\bigwedge_{\iota \in I} \mu_\iota$ and therefore

$\bigwedge_{\iota \in I} \mu_\iota = 0$. □

Corollary 3.4.5 *Let N be a fundamental solid subspace of M, let $\underline{\underline{P}}$ be an upper directed family of bands of N, let P_0 be the band of N generated by $\bigcup_{P \in \underline{\underline{P}}} P$, let $\underline{\underline{F}}$ be the section filter of $\underline{\underline{N}}$ with respect to the inclusion relation, and let $x \in P_0$; for any $P \in \underline{\underline{P}}$ let x_P be the component of x on P. Then the map*

$$\underline{\underline{P}} \rightarrow (N, N^\pi), \quad P \mapsto x_P$$

converges along $\underline{\underline{F}}$ to x.

Assume first x positive. Then $(x_P)_{P \in \underline{\underline{P}}}$ is an upper directed family in N whose supremum is x and the assertion follows from the preceding proposition.

Let now x be arbitrary and for any $P \in \underline{\underline{P}}$ let x'_P, x''_P be the components of x_+, x_- on P respectively. By the above remark the maps

$$\underline{\underline{P}} \rightarrow (N, N^\pi), \quad P \mapsto x'_P,$$

$$\underline{\underline{P}} \rightarrow (N, N^\pi), \quad P \mapsto x''_P$$

converge along $\underline{\underline{F}}$ to x_+ and x_- respectively and the assertion follows immediately. □

Corollary 3.4.6 *Let N be a fundamental solid subspace of M, let $(N_\iota)_{\iota \in I}$ be a family of bands of N such that N is direct sum as vector lattice of this family, let $\mu \in N$, and let $(\mu_\iota)_{\iota \in I}$ be the family of components of μ with respect to $(N_\iota)_{\iota \in I}$. Then $(\mu_\iota)_{\iota \in I}$ is summable and its sum is μ in (N, N^π).*

This corollary follows immediately from the preceding one. □

Proposition 3.4.7 *Let* N *be a fundamental solid subspace of* M, *let* F *be a subset of* N^{π} *generating it as band, let* P *be a subset of* N, *let* B *be the band of* N *generated by* P, *let* $\underline{\underline{S}}$ *be a ring of sets generating the* δ*-ring* $\underline{\underline{R}}$, *let* G *be the set of step functions on* X *with respect to* $\underline{\underline{S}}$, *and let* ν *be an element of* B_+. *We set*

$$P' := \{ \sum_{\iota\in I} g_\iota \cdot \mu_\iota \mid (g_\iota, \mu_\iota)_{\iota\in I} \text{ finite family in } G_+ \times P\} ,$$

$$P'' := \{ \sum_{\iota\in I} g_\iota \cdot \mu_\iota \mid (g_\iota, \mu_\iota)_{\iota\in I} \text{ finite family in } G \times P\} ,$$

$$P'_0 := \{\lambda\in P' \mid \lambda\leq\nu\}$$

and denote by $\bar{P}'$, $\bar{P}''$, $\bar{P}'_0$ *the closures of* P' , P'' , P'_0 *in* (N,F) *respectively. Then*

a) $P\subset N_+ \, , \, \underline{\underline{R}} = \underline{\underline{S}} \implies \nu\in\bar{P}'_0$;

b) $P\subset N_+ \implies B_+ = \bar{P}'$;

c) $B = P''$.

By Proposition 3.4.2 we may assume F upper directed. Since F generates N^{π} as band, (N,F) is Hausdorff and therefore B and B_+ are closed. Hence $P''\subset B$ and

$$P\subset N_+ \implies \bar{P}'\subset B_+ .$$

a) & b) Assume first P finite. Then there exists a finite family $(\nu_\mu)_{\mu\in P}$ in N_+ such that $\nu = \sum_{\mu\in P} \nu_\mu$ and such that $\nu_\mu \ll \mu$ for any $\mu\in P$. Let $\xi\in F$. Then $|\xi|\cdot\nu_\mu \ll |\xi|\cdot\mu\in M_b$ for any $\mu\in P$ (Theorem 3.2.2d)). By Radon-Nikodym theorem there exists $(f_\mu)_{\mu\in P}\in \prod_{\mu\in P} L^1(|\xi|\cdot\mu)_+$ such that $|\xi|\cdot\nu_\mu = f_\mu\cdot(|\xi|\cdot\mu)$ for any $\mu\in P$. Let $\varepsilon>0$. There exists a family $(g_\mu)_{\mu\in P}$ in G_+ such that

$$\int |f_\mu - g_\mu| d(|\xi|\cdot\mu) < \frac{\varepsilon}{n}$$

for any $\mu\in P$, where n denotes the cardinal number of P . Moreover if $\underline{\underline{R}} = \underline{\underline{S}}$ we may assume $g_\mu \leq f_\mu$ for any $\mu\in P$. We get (Theorems 3.1.7 b) and 3.2.2 d) f))

$$q_\xi(\nu_\mu - g_\mu\cdot\mu) = \int |\xi| d|\nu_\mu - g_\mu\cdot\mu| = \int 1_X d(|\xi|\nu_\mu - g_\mu\cdot(|\xi|\cdot\mu)) =$$

$$= \int |f_\mu - g_\mu| d(|\xi|\cdot\mu) < \frac{\varepsilon}{n}$$

for any $\mu \in P$ and therefore

$$q_\xi(\nu - \sum_{\mu\in P} g_\mu\cdot\mu) \leq \sum_{\mu\in P} q_\xi(\nu_\mu - g_\mu\cdot\mu) < \varepsilon .$$

Hence $\nu \in \bar{P}'$ and if $\underline{\underline{R}} = \underline{\underline{S}}$ then $\nu \in \bar{P}'_o$.

Let now P be arbitrary and let $\underline{\underline{P}}$ be the set of bands N generated by finite subsets of P . For any $P_o \in \underline{\underline{P}}$ we denote by ν_{P_o} the component of ν on P_o . By the above proof $\nu_{P_o} \in \bar{P}'$ and if $\underline{\underline{R}} = \underline{\underline{S}}$ then $\nu_{P_o} \in \bar{P}'_o$. By Corollary 3.4.5 ν belongs to the closure of $\{\nu_{P_o} \mid P_o \in \underline{\underline{P}}\}$ in (N,F). Hence $\nu \in \bar{P}'$ and if $\underline{\underline{R}} = \underline{\underline{S}}$ then $\nu \in \bar{P}'_o$.

c) Assume first $P \subset N_+$. Since $P'' = P' - P'$ we get

$$B = B_+ - B_+ = \bar{P}' - \bar{P}' \subset \overline{P' - P'} = \bar{P}'' .$$

Let now P be arbitrary and let B_1 , B_2 be the bands of N generated by $P \cap N_+$ and $P \cap N_-$ respectively. Let $\lambda \in B$. Then there exists $(\lambda_1, \lambda_2) \in B_1 \times B_2$ with $\lambda = \lambda_1 + \lambda_2$. By the above remark $\lambda \in \bar{P}''$. □

<u>*Theorem 3.4.8*</u>

a) *For any* $\xi \in M^\pi$ *and for any* η, $\eta' \in M^\pi_c$ *we have*

$$p_\xi(\eta\eta') \leq p_\xi(\eta) p_\xi(\eta') ;$$

b) (M^π_c, M^π) *is a topologically and orderly complete Hausdorff locally convex lattice and locally convex algebra ;*

c) M^π_b *is a dense set of* (M^π_c, M^π) ;

d) *if there exists* $\xi \in M^\pi$ *such that* p_ξ *is a norm then* $M^\pi_c = M^\pi_b$.

a), b), c) follow from Theorem 2.3.8 and Propositions 1.6.5 and 3.1.1 c).

d) Let (Y,u,v) be a representation of (X,M) (Theorem 2.3.8). For any $\eta \in M^\pi_c$ we have (Proposition 3.1.1 b))

$$p_\xi(\eta) = \sup_{x \in \mathrm{Supp}(\hat{u}\xi)} |(\hat{u}\eta)(x)| .$$

Since p_ξ is a norm we have

$$\mathrm{Supp}\,(\hat{u}\xi) = Y .$$

By Propositions 3.1.1 b) c) and 1.6.2 $M_c^\pi = M_b^\pi$. □

Remark. The following example will show that M^π may be neither closed nor dense in (M_c^π, M^π) even if (M_c^π, M^π) is normable. We set $X := \mathbb{N}$,

$$\underline{\underline{R}} := \{A \subset \mathbb{N} \mid \sum_{n\in A} \frac{1}{n} < \infty\}$$

and take as M the set of measures on $\underline{\underline{R}}$. Then $\dot{i}_X$ does not belong to the closure of M^π in (M_c^π, M^π) while

$$M_c \longrightarrow \mathbb{R} , \mu \longmapsto \sum_{n\in\mathbb{N}} \mu(\{n\})(\log n)^{-1}$$

belongs to the closure of M^π in (M_c^π, M^π) but not to M^π . In this example $M_b \neq M_c$.

Theorem 3.4.9 *If* F *denotes the set* $\{\dot{i}_A \mid A\in\underline{\underline{R}}\}$ *then any bounded set of* (M_c^π, F) *is relatively compact with respect to the* $\sigma(M_c^\pi, M_c)$*-topology.*

Let G be a bounded set of positive elements of (M_c^π, F) , let $\underline{\underline{F}}$ be an ultrafilter on G , and let ξ be the map

$$M_c \longrightarrow \mathbb{R} , \mu \longmapsto \lim_{\eta,\underline{\underline{F}}} \int \eta \, d\mu .$$

Then $\xi\in M_c^+$ and therefore by Theorem 3.3.2 $\xi\in M_c^\pi$. Hence G is a relatively compact set of M_c^π with respect to $\sigma(M_c^\pi, M_c)$.

Let now G be an arbitrary bounded set of (M_c^π, F) . Since the sets

$$G' := \{\xi \vee 0 \mid \xi\in G\} , \quad G'' := \{(-\xi) \vee 0 \mid \xi\in G\}$$

are bounded sets of (M_c^π, F), it follows from the above considerations that they are relatively compact with respect to $\sigma(M_c^\pi, M_c)$. The map

$$M_c^\pi \times M_c^\pi \longrightarrow M_c^\pi , \ (\xi,\eta) \longmapsto \xi-\eta$$

being continuous with respect to $\sigma(M_c^\pi, M_c)$ the set $G'-G"$ is relatively compact with respect to $\sigma(M_c^\pi, M_c)$. From $G \subset G'-G"$ we see that G is relatively compact with respect to $\sigma(M_c^\pi, M_c)$. □

Theorem 3.4.10 *Let* G *be a solid subspace of* M_c^π, *let* F *be a subset of* M^π *containing* $\{1_A | A \in \underline{\underline{R}}\}$, *let* G' *be the dual space of* (G,F), *let* (Y,u,v) *be a representation of* (X,M), *let* Y^* *be the Stone-Čech compactification of* Y, *and for any* $f \in C_\infty(Y)$ *let* f^* *denote the continuous extention of* f *to* Y^* . *Let further* Y_0 *be the locally compact space* $\bigcup_{\xi \in F} \operatorname{Supp}(\hat{u}\xi)^*$, *let* w *be the map*

$$C(Y_0) \longrightarrow C(Y) , \ f \longmapsto f|Y ,$$

and let M_0 *be the set of Radon real measures* λ *on* Y_0 *with compact support such that*

$$\bigcap_{\eta \in G} (\{(\hat{u}\eta)^* = 0\} \cap Y_0)$$

is a λ*-null set. Then:*

a) w *is an isomorphism of vector lattices ;*

b) *for any* $\lambda \in M_0$ *we have* $C(Y_0) \subset L^1(\lambda)$; *we denote by* λ' *the map (Proposition 3.1.1 c))*

$$G \longrightarrow \mathbb{R} , \ \eta \longmapsto \int w^{-1} \hat{u}\eta d\lambda = \int (\hat{u}\eta)^* | Y_0 d\lambda \ ;$$

c) *for any* $\lambda \in M_0$ *we have* $\lambda' \in G'$;

d) G' *is a solid subspace of* G^+ *and the map*

$$M_0 \longrightarrow G', \lambda \longmapsto \lambda'$$

is an isomorphism of vector lattices.

a) follows from Proposition 3.1.1 c) and 1.6.2.

b) is trivial.

c) Since Supp λ is compact and $(\operatorname{Supp}(\hat{u}\xi)^*)_{\xi \in F}$ is an open covering of Y_o it follows that there exists a finite subset F_o of F such that

$$\mathrm{Supp}\,\lambda \subset \bigcup_{\xi\in F_o} (\mathrm{Supp}(\hat{u}\xi)^*)$$

and we get

$$|\lambda'(\eta)| \leqslant \|\lambda\| \sup_{\xi\in F_o} p_\xi(\eta)$$

for any $\eta\in G$. Hence $\lambda'\in G'$.

d) The only non-trivial assertion is the surjectivity of the map. Let $\varphi\in G'$. Let us endow $C(Y_o)$ with the topology of compact convergence. By Hahn-Banach theorem there exists a continuous linear form ψ on $C(Y_o)$ such that

$$\psi \circ w^{-1} \circ (\hat{u}|G) = \varphi \ .$$

There exists a Radon real measure λ_o on Y_o with compact support such that

$$\psi(f) = \int f d\lambda_o$$

for any $f\in C(Y_o)$. M_o is a band of the vector lattice of Radon real measures on Y_o with compact support. We denote by λ the component of λ_o on M_o . Then

$$\lambda'(\eta) = \int w^{-1}\hat{u}\eta d\lambda =$$

$$= \int w^{-1}\hat{u}\eta d\lambda_o = \psi(w^{-1}\hat{u}\eta) = \varphi(\eta)$$

for any $\eta\in G$ and therefore $\lambda' = \varphi$. □

<u>*Theorem 3.4.11*</u> *Let* I *be a countable set and let* φ *be a continuous real function on* $\mathbb{R}^I$ *. Then:*

a) there exists a unique map $\hat{\varphi}:(M_c^\pi)^I \longrightarrow M_c^\pi$ *such that for any representation* (Y,u,v) *of* (X,M)*, for any family* $(\xi_\iota)_{\iota\in I}$ *in* M_c^π *, and for any* $y\in Y$ *we have*

$$(\hat{u}\hat{\varphi}((\xi_\iota)_{\iota\in I}))(y) = \varphi(((\hat{u}\xi_\iota)(y))_{\iota\in I}) \ ;$$

b) if ψ *denotes a Borel measurable real function on* $\bar{\mathbb{R}}^I$ *whose restriction to* $\mathbb{R}^I$ *is equal to* φ *then* $\hat{\varphi}$ *is the restriction to* $(M_c^\pi)^I$ *of the map* $\hat{\psi}$ *defined in Theorem 3.1.5 ;*

c) for any $(\xi_\iota)_{\iota\in I}\in(M_c^\pi)^I$, *for any* $\xi\in M^\pi$, *and for any* $\varepsilon>0$ *there exists a finite subset* J *of* I *and a* $\delta>0$ *such that*

$$p_\xi(\hat{\varphi}((\xi_\iota)_{\iota\in I}) - \hat{\varphi}((\eta_\iota)_{\iota\in I})) < \varepsilon$$

for any $(\eta_\iota)_{\iota\in I}\in(M_c^\pi)^I$ *with*

$$\sup_{\iota\in J} p_\xi(\xi_\iota-\eta_\iota) < \delta .$$

d) if F *denotes a subset of* M^π *then* $\hat{\varphi}$ *is continuous with respect to the topology of* (M_c^π, F) *;*

e) if φ *is uniformly continuous and* F *denotes a subset of* M^π *then* $\hat{\varphi}$ *is uniformly continuous with respect to the uniformity of* (M_c^π, F) *.*

a) By Proposition 3.1.1 c) $(\hat{u}\xi_\iota)_{\iota\in I}\in(C(Y))^I$ and therefore the map

$$Y \longrightarrow \mathbb{R}^I , \; y \longmapsto ((\hat{u}\xi_\iota)(y))_{\iota\in I}$$

is continuous. We deduce that the map

$$Y \longrightarrow \mathbb{R} , \; y \longrightarrow \varphi(((\hat{u}\xi_\iota)(y))_{\iota\in I})$$

is continuous. By Proposition 3.1.1 b) c) there exists a unique element $\hat{\varphi}((\xi_\iota)_{\iota\in I})\in M_c^\pi$ such that

$$(\hat{u}\hat{\varphi}((\xi_\iota)_{\iota\in I}))(y) = \varphi(((\hat{u}\xi_\iota)(y))_{\iota\in I})$$

for any $y\in Y$.

b) is obvious.

c) By Proposition 3.1.1 c) and Propisition 1.6.2 there exists a family $(K_\iota)_{\iota\in I}$ of compact sets of $\mathbb{R}$ such that $(\hat{u}\xi_\iota)(y)\in K_\iota$ for any $y\in \mathrm{Supp}(\hat{u}\xi)$ and any $\iota\in I$. The set $\prod_{\iota\in I} K_\iota$ being a compact set of

$\mathbb{R}^I$ the restriction of φ to it is uniformly continuous. Hence there exists a finite subset J of I and a $\delta>0$ such that

$$|\varphi((\alpha_\iota)_{\iota\in I}) - \varphi((\beta_\iota)_{\iota\in I})| < \varepsilon$$

for any

$$(\alpha_\iota)_{\iota\in I}\ ,\ (\beta_\iota)_{\iota\in I} \in \prod_{\iota\in I} K_\iota$$

with

$$\sup_{\iota\in J} |\alpha_\iota - \beta_\iota| < \delta .$$

Let $(\eta_\iota)_{\iota\in I} \in (M_c^\pi)^I$ with

$$\sup_{\iota\in J} p_\xi(\xi_\iota - \eta_\iota) < \delta .$$

Then $|\hat{u}\xi_\iota - \hat{u}\eta_\iota| < \delta$ on $\mathrm{Supp}(\hat{u}\xi)$ for any $\iota\in J$. We get by the above considerations

$$|\hat{u}\hat{\varphi}((\xi_\iota)_{\iota\in I}) - \hat{u}\hat{\varphi}((\eta_\iota)_{\iota\in I})| < \varepsilon$$

on $\mathrm{Supp}\,(\hat{u}\xi)$ and therefore

$$p_\xi(\hat{\varphi}((\xi_\iota)_{\iota\in I}) - \hat{\varphi}((\eta_\iota)_{\iota\in I})) < \varepsilon .$$

d) follows immediately from c).

e) Let $\xi\in F$ and let ε be a strictly positive real number. There exist a finite subset J of I and a strictly positive real number δ such that

$$|\varphi((\alpha_\iota)_{\iota\in I}) - \varphi((\beta_\iota)_{\iota\in I})| < \varepsilon$$

for any $(\alpha_\iota)_{\iota\in I}$, $(\beta_\iota)_{\iota\in I} \in \mathbb{R}^I$ such that

$$\sup_{\iota\in J} |\alpha_\iota - \beta_\iota| < \delta .$$

Let $(\xi_\iota)_{\iota\in I}$, $(\eta_\iota)_{\iota\in I} \in M_c^\pi$ such that

$$\sup_{\iota \in J} p_\xi(\xi_\iota - \eta_\iota) < \delta .$$

Then $|\hat{u}\xi_\iota - \hat{u}\eta_\iota| < \delta$ on Supp $(\hat{u}\xi)$ for any $\iota \in I$. We get

$$|\hat{u}\hat{\varphi}((\xi_\iota)_{\iota \in I}) - \hat{u}\hat{\varphi}((\eta_\iota)_{\iota \in I})| < \varepsilon$$

on Supp $(\hat{u}\xi)$ and therefore

$$p_\xi(\hat{\varphi}((\xi_\iota)_{\iota \in I}) - \hat{\varphi}((\eta_\iota)_{\iota \in I})) < \varepsilon . \square$$

Theorem 3.4.12 *Let* I *be a countable set, let* F *be a subset of* M^π*, and let* Ψ*, (*Ψ'*) be the set of continuous maps* $\mathbb{R}^I \longrightarrow \mathbb{R}$*,* $((M_c^\pi, F)^I \longrightarrow (M_c^\pi, F))$ *endowed with the topology of uniform convergence on the bounded sets. The map*

$$\Psi \longrightarrow \Psi', \ \varphi \longmapsto \hat{\varphi}$$

defined in Theorem 3.4.11 a) is continuous.

Let $\varphi \in \Psi$, let A be a bounded set of $(M_c^\pi)^I$, let $\xi \in M^\pi$ and let ε be a strictly positive real number. Then for any $\iota_0 \in I$

$$\bar{\alpha}_{\iota_0} := \sup \{p_\xi(\xi_{\iota_0}) \mid (\xi_\iota)_{\iota \in I} \in A\} < \infty .$$

Let $\psi \in \Psi$ such that

$$|\varphi((\alpha_\iota)_{\iota \in I}) - \psi((\alpha_\iota)_{\iota \in I})| < \varepsilon$$

for any $(\alpha_\iota)_{\iota \in I} \in \mathbb{R}^I$ with $|\alpha_\iota| \leq \bar{\alpha}_\iota$ for any $\iota \in I$.

Let (Y,u,v) be a representation of (X,M) (Theorem 2.3.8) and let $(\xi_\iota)_{\iota \in I} \in A$. Then $|\hat{u}\xi_\iota| \leq \bar{\alpha}_\iota$ on Supp $(\hat{u}\xi)$ for any $\iota \in I$ and therefore

$$|\varphi((\hat{u}\xi_\iota)_{\iota \in I} - \psi((\hat{u}\xi_\iota)_{\iota \in I})| < \varepsilon$$

on Supp $(\hat{u}\xi)$. We get

$$|\hat{u}\hat{\varphi}((\xi_\iota)_{\iota \in I}) - \hat{u}\hat{\psi}((\xi_\iota)_{\iota \in I})| < \varepsilon$$

on Supp $(\hat{u}\xi)$ and therefore

$$p_\xi(\hat{\varphi}((\xi_\iota)_{\iota\in I}) - \hat{\psi}((\xi_\iota)_{\iota\in I})) < \varepsilon .$$

Hence the map

$$\Psi \longrightarrow \Psi', \quad \varphi \longmapsto \hat{\varphi}$$

is continuous. □

Proposition 3.4.13 *Let* I *be a finite set and let* φ *be a locally bounded, Borel measurable real function on* $\mathbb{R}^I$. *Then for any family* $(f_\iota)_{\iota\in I}$ *in* L_c *the function*

$$f: X \longrightarrow \mathbb{R}, \; x \longmapsto \varphi((f_\iota(x))_{\iota\in I})$$

belongs to L_c.

The assertion is obvious if φ takes only a finite number of values. If φ is positive we may approximate it by an increasing sequence of step functions. Since f is bounded on any $A\in\underline{\underline{R}}$ we easily deduce that $f\in L_c$ in this case. For a general φ the assertion follows by representing φ as a difference of two positive, locally bounded, Borel measurable functions on $\mathbb{R}^I$. □

Theorem 3.4.14 *Let* I *be a finite set, let* φ *be a locally bounded, Borel measurable real function on* $\mathbb{R}^I$, *and for any* $(f_\iota)_{\iota\in I}\in(L_c)^I$ *let* $\varphi((f_\iota)_{\iota\in I})$ *be the function*

$$X \longrightarrow \mathbb{R}, \; x \longmapsto \varphi((f_\iota(x))_{\iota\in I}) .$$

Then:

a) there exists a unique map $\hat{\varphi}: (M_c^\pi)^I \longrightarrow M_c^\pi$ *possessing the following property: for any* $\mu\in M_c$ *and for any* $(\xi_\iota)_{\iota\in I}\in(M_c^\pi)^I$ $(f_\iota)_{\iota\in I}\in(L_c)^I$ *such that* $\xi_\iota\cdot\mu = f_\iota\cdot\mu$ *for any* $\iota\in I$ *we have*

$$\hat{\varphi}((\xi_\iota)_{\iota\in I})\cdot\mu = \varphi((f_\iota)_{\iota\in I})\cdot\mu \;;$$

b) for any $(f_\iota)_{\iota\in I}\in(L_c)^I$ *we have*

$$\hat{\varphi}((\dot{f}_\iota)_{\iota\in I}) = \widehat{\varphi((f_\iota)_{\iota\in I})}^{\,\cdot} \ ;$$

c) for any $\mu\in M$ and for any $(\xi_\iota)_{\iota\in I}$, $(\eta_\iota)_{\iota\in I}\in(M_c^\pi)^I$ such that $\xi_\iota\cdot\mu = \eta_\iota\cdot\mu$ for any $\iota\in I$ we have

$$\hat{\varphi}((\xi_\iota)_{\iota\in I})\cdot\mu = \hat{\varphi}((\eta_\iota)_{\iota\in I})\cdot\mu \ ;$$

d) if ψ is a Borel measurable real function on $\overline{\mathbb{R}}^I$ equal to φ on $\mathbb{R}^I$ then $\hat{\psi}$ defined in Theorem 3.1.5 is equal to $\hat{\varphi}$ on $(M_c^\pi)^I$;

e) if φ is continuous then $\hat{\varphi}$ coincides with $\hat{\varphi}$ defined in Theorem 3.4.11.

By Proposition 3.4.13 $\varphi((f_\iota)_{\iota\in I})\in L_c$ for any $(f_\iota)_{\iota\in I}\in(L_c)^I$. Let ψ be a Borel measurable real function on $\overline{\mathbb{R}}^I$ equal to φ on $\mathbb{R}^I$ and let (Y,u,v) be a representation of (X,M) (Theorem 2.3.8). Then for any $(\xi_\iota)_{\iota\in I}\in(M_c^\pi)^I$ we have $(\hat{u}\xi_\iota)_{\iota\in I}\in C(Y)^I$ (Proposition 3.1.1 c)) and therefore $\psi((\hat{u}\xi_\iota)_{\iota\in I})$ is locally bounded. Hence $\hat{\psi}((\xi_\iota)_{\iota\in I})\in M_c^\pi$ (Proposition 3.1.1 b) c)). We set

$$\hat{\varphi} : (M_c^\pi)^I \longrightarrow M_c^\pi \ , \ (\xi_\iota)_{\iota\in I} \longmapsto \hat{\psi}((\xi_\iota)_{\iota\in I}) \ .$$

b) By Theorem 3.1.5 we have

$$\hat{\varphi}((\dot{f}_\iota)_{\iota\in I}) = \hat{\psi}((\dot{f}_\iota)_{\iota\in I}) = \widehat{\psi((f_\iota)_{\iota\in I})}^{\,\cdot} = \widehat{\varphi((f_\iota)_{\iota\in I})}^{\,\cdot} \ .$$

c) Let $\mu\in M$ and let $(\xi_\iota)_{\iota\in I}$, $(\eta_\iota)_{\iota\in I}\in(M_c^\pi)^I$ such that $\xi_\iota\cdot\mu = \eta_\iota\cdot\mu$ for any $\iota\in I$. We have (Proposition 3.1.1 e))

$$(\hat{u}\xi_\iota)\cdot(v\mu) = v(\xi_\iota\cdot\mu) = v(\eta_\iota\cdot\mu) = (\hat{u}\eta_\iota)\cdot(v\mu)$$

and therefore $\hat{u}\xi_\iota = \hat{u}\eta_\iota$ on $\mathrm{Supp}\,(v\mu)$ for any $\iota\in I$. We deduce (Proposition 3.1.1 b) e) and Theorem 3.1.5)

$$v(\hat{\varphi}((\xi_\iota)_{\iota\in I})\cdot\mu) = v(\hat{\psi}((\xi_\iota)_{\iota\in I})\cdot\mu) = (\hat{u}\hat{\psi}((\xi_\iota)_{\iota\in I}))\cdot(v\mu) =$$

$$= (\psi((\hat{u}\xi_\iota)_{\iota\in I}))\cdot(v\mu) = (\psi((\hat{u}\eta_\iota)_{\iota\in I}))\cdot(v\mu) =$$

$$= v(\hat{\varphi}((\xi_\iota)_{\iota\in I})\cdot\mu) \ ,$$

$$\hat{\varphi}((\xi_\iota)_{\iota\in I})\cdot\mu = \hat{\varphi}((\eta_\iota)_{\iota\in I})\cdot\mu \ .$$

a) Let $\mu\in M_c$ and let $(\xi_\iota)_{\iota\in I}\in(M_c^\pi)^I$, $(f_\iota)_{\iota\in I}\in(L_c)^I$ such that $\xi_\iota\cdot\mu = f_\iota\cdot\mu$ for any $\iota\in I$. We have $(\dot{f}_\iota)_{\iota\in I}\in(M_c^\pi)^I$ (Proposition 1.5.5 c)) and $\xi_\iota\cdot\mu = \dot{f}_\iota\cdot\mu$ for any $\iota\in I$. By c) we get

$$\hat{\varphi}((\xi_\iota)_{\iota\in I})\cdot\mu = \hat{\varphi}((\dot{f}_\iota)_{\iota\in I})\cdot\mu$$

and therefore by b)

$$\hat{\varphi}((\xi_\iota)_{\iota\in I})\cdot\mu = \dot{\widehat{\varphi((f_\iota)_{\iota\in I})}}\cdot\mu = \varphi((f_\iota)_{\iota\in I})\cdot\mu \ .$$

In order to prove the unicity let φ^* be a map $(M_c^\pi)^I \longrightarrow M_c^\pi$ possessing the indicated property and let $(\xi_\iota)_{\iota\in I}\in(M_c^\pi)^I$. Let further $\mu\in M_c$. By Propositions 1.4.1, 2.4.4 c) and 3.1.1 c) e) there exists $(f_\iota)_{\iota\in I}\in(L_c)^I$ such that $uf_\iota = \hat{u}\xi_\iota$ on Supp $(v\mu)$ for any $\iota\in I$ and we get

$$\varphi^*((\xi_\iota)_{\iota\in I})\cdot\mu = \varphi((f_\iota)_{\iota\in I})\cdot\mu = \hat{\varphi}((\xi_\iota)_{\iota\in I})\cdot\mu \ ,$$

$$(\hat{u}\varphi^*((\xi_\iota)_{\iota\in I}))\cdot(v\mu) = (\hat{u}\hat{\varphi}((\xi_\iota)_{\iota\in I}))\cdot(v\mu) \ ,$$

$$\hat{u}\varphi^*((\xi_\iota)_{\iota\in I}) = \hat{u}\hat{\varphi}((\xi_\iota)_{\iota\in I}) \text{ on Supp } (v\mu) \ .$$

By Propositions 2.4.4 a) and 3.1.1 b) $\varphi^*((\xi_\iota)_{\iota\in I}) = \hat{\varphi}((\xi_\iota)_{\iota\in I})$.

d) is obvious from the definition of $\hat{\varphi}$ and Proposition 3.1.1 c)

e) follows from d) and Theorem 3.4.11 b). □

Theorem 3.4.15 *Let* F *be a subset of* M^π. (M,F) *endowed with the action* (Proposition 3.2.1)

$$M_c^\pi\times M \longrightarrow M \ , \ (\xi,\mu) \longmapsto \xi\cdot\mu$$

is a topological M_c^π*-module such that:*

a) for any $\xi\in M_{c+}^\pi$ *and for any upper bounded nonempty family* $(\mu_\iota)_{\iota\in I}$ *in* M *the family* $(\xi\cdot\mu_\iota)_{\iota\in I}$ *is upper bounded and*

$$\xi\cdot(\bigvee_{\iota\in I}\mu_\iota) = \bigvee_{\iota\in I}(\xi\cdot\mu_\iota) \quad ;$$

b) *for any* $\mu\in M_+$ *and for any upper bounded nonempty family* $(\xi_\iota)_{\iota\in I}$ *in* M_c^π *the family* $(\xi_\iota\cdot\mu)_{\iota\in I}$ *is upper bounded and*

$$(\bigvee_{\iota\in I}\xi_\iota)\cdot\mu = \bigvee_{\iota\in I}(\xi_\iota\cdot\mu) \quad ;$$

c) M_c *is an* M_c^π*-submodule of* M *and* M_b *is an* M_b^π*-submodule of* M ;

d) *for any* $\mu\in M$ *there exist* $\xi,\eta\in(M_b^\pi)_+$ *such that*

$$\xi^2 = \xi,\ \eta^2 = \eta,\ \xi\cdot\eta = \xi\wedge\eta = 0 \ ,$$

$$\xi\cdot\mu = \mu_+ \ ,\ \eta\cdot\mu = -\mu_- \ ,\ \ (\xi-\eta)\cdot\mu = |\mu| \ ;$$

e) *for any* $(\xi,\mu)\in M_c^\pi\times M$ *we have* $\xi\cdot\mu<<\mu$ *and* $|\xi\cdot\mu| = |\xi|\cdot|\mu|$;

f) *any* $\xi\in M_c^\pi$ $(\mu\in M)$ *is positive if* $\xi\cdot\mu \geq 0$ *for any* $\mu\in M_{c+}^\pi$ $(\xi\in(M^\pi)_+)$;

g) *we have* $q_\xi(\eta\cdot\mu) \leq p_\xi(\eta)q_\xi(\mu)$ *for any* $(\xi,\eta,\mu)\in M^\pi\times M_c^\pi\times M$.

The assertions follow from Theorem 2.3.8, Proposition 3.1.1, Theorem 3.1.7, Theorem 3.4.3 and Theorem 3.4.8. □

5. Spaces of operators

Proposition 3.5.1 *Let* φ *be a linear map* $M \longrightarrow M$ *such that* $\varphi\mu<<\mu$ *for any* $\mu\in M$. *Then:*

a) *for any* $\mu\in M$ *there exists a unique* $\xi_\mu\in\hat{L}^1_{loc}(\mu)$ *such that* $\varphi\mu = \xi_\mu\cdot\mu$;

b) $\xi_{\alpha\mu} = \xi_\mu$ *for any* $(\alpha,\mu)\in\mathbb{R}\times M$ *with* $\alpha\neq 0$;

c) $\xi_{\mu+\nu} = \xi_\mu+\xi_\nu$ *for any* $\mu,\nu\in M$ *with* $|\mu|\wedge|\nu| = 0$

d) $\xi_{\eta\cdot\mu} = \eta\xi_\mu$ *for any* $\eta\in M_c^\pi$ *with* $\eta^2 = \eta$.

a) follows immediately from Theorem 3.2.2 c) .

b) We have $\hat{L}^1_{loc}(\mu) = \hat{L}^1_{loc}(\alpha\mu)$ and

$$\xi_{\alpha\mu}\cdot\mu = \frac{1}{\alpha}\,\xi_{\alpha\mu}\cdot(\alpha\mu) = \frac{1}{\alpha}\,\varphi(\alpha\mu) = \varphi\mu = \xi_\mu\cdot\mu$$

and therefore $\xi_{\alpha\mu} = \xi_\mu$ (Theorem 3.2.2 c)).

c) By Corollary 3.1.4 $\hat{L}^\perp(\mu)\cap\hat{L}^\perp(\nu) = \{0\}$ and therefore $\xi_\mu\in\hat{L}(\nu)$, $\xi_\nu\in\hat{L}(\mu)$ (Corollary 3.1.3 c)). We get (Corollary 3.1.3 a) Proposition 3.2.6 b))

$$\xi_\mu + \xi_\nu\in\hat{L}^1_{loc}(\mu)\cap\hat{L}^1_{loc}(\nu)\subset\hat{L}^1_{loc}(\mu+\nu)$$

and (Theorem 3.2.2 b) Proposition 3.2.6 b))

$$(\xi_\mu+\xi_\nu)\cdot(\mu+\nu) = \xi_\mu\cdot\mu+\xi_\nu\cdot\nu=\varphi\mu+\varphi\nu=\varphi(\mu+\nu) = \xi_{\mu+\nu}\cdot(\mu+\nu) .$$

By Proposition 3.2.6 d) $\xi_\mu+\xi_\nu\ \hat{L}^1(\mu+\nu)$ and therefore (Theorem 3.2.2 c)) $\xi_\mu+\xi_\nu = \xi_{\mu+\nu}$.

d) By Theorem 3.1.7 i) $\eta\wedge(\dot{1}_X-\eta) = 0$ and therefore (Theorem 3.2.2 b), f))

$$|\eta\cdot\mu|\wedge|(\dot{1}_X-\eta)\cdot\mu| = (\eta\cdot|\mu|)\wedge((\dot{1}_X-\eta)\cdot|\mu|) = 0 .$$

By c) we deduce

$$\xi_\mu = \xi_{\eta\cdot\mu} + \xi_{(\dot{1}_X-\eta)\cdot\mu} .$$

We have (Theorem 3.2.2 b), d), Proposition 3.2.1)

$$(\eta\xi_{\eta\cdot\mu})\cdot(\eta\cdot\mu) = \xi_{\eta\cdot\mu}\cdot(\eta^2\cdot\mu) = \xi_{\eta\cdot\mu}\cdot(\eta\cdot\mu) ,$$

$$(\eta\xi_{(\dot{1}_X-\eta)\cdot\mu})\cdot((\dot{1}_X-\eta)\cdot\mu) = \xi_{(\dot{1}_X-\eta)\cdot\mu}((\eta(\dot{1}_X-\eta))\cdot\mu) = 0$$

and therefore (Theorem 3.2.2 c), Theorem 3.1.7)

$$\eta\xi_{\eta\cdot\mu} = \xi_{\eta\cdot\mu} , \quad \eta\xi_{(\dot{1}_X-\eta)\cdot\mu} = 0 ,$$

$$\eta\xi_{\mu} = \eta\xi_{\eta\cdot\mu} + \eta\xi_{(\dot{1}_X-\eta)\cdot\mu} = \xi_{\eta\cdot\mu} . \quad \square$$

Definition 3.5.2 Let $L_d(M)$ be the set of linear maps $\varphi : M \longrightarrow M$ such that for any $\xi\in M^\pi$ there exists $\alpha\in\mathbb{R}_+$ with $q_\xi \circ \varphi \leqslant \alpha q_\xi$. For any $\xi\in M^\pi$ we denote by r_ξ the map

$$L_d(M) \longrightarrow \mathbb{R} , \ \varphi \longmapsto \inf\{\alpha\in\mathbb{R}_+ | q_\xi \circ\varphi \leqslant \alpha q_\xi\} .$$

For any $\varphi, \psi\in L_d(M)$ we set

$$\varphi \leqslant \psi \ :\Longleftrightarrow (\forall\mu\in M_+ \Longrightarrow \varphi\mu \leqslant \psi\mu) .$$

It is obvious that $L_d(M)$ is a subalgebra of the algebra of continuous linear maps $M \longrightarrow M$ and that $\leqslant$ is an order relation on $L_d(M)$.

Theorem 3.5.3 Let φ be a linear map $M \longrightarrow M$. The following assertions are equivalent:

a) $\varphi\in L_d(M)$;

b) $\varphi\mu << \mu$ *for any* $\mu\in M$;

c) *there exists* $\eta\in M^\pi_c$ *such that* $\varphi\mu = \eta\cdot\mu$ *for any* $\mu\in M$.

a $\Rightarrow$ b . Let $A\in\underline{\underline{R}}$ be a μ-null set. We put $\xi := \dot{1}_A$. Then $\xi\in M^\pi$ and $\xi\cdot\mu = 0$. We get successively $q_\xi(\mu) = 0$, $q_\xi(\varphi\mu) = 0$, $\xi\cdot|\varphi\mu| = 0$, A is a $\varphi\mu$-null set. Hence $\varphi\mu << \mu$.

b $\Rightarrow$ a . It is obvious that for any $\mu\in M$ and for any band N of M the component of $\varphi\mu$ on N is equal to $\varphi\nu$, where ν denotes the component of μ on N . Let (Y,u,v) be a representation of (X,M) (Theorem 2.3.8) and let $\xi\in M^\pi$. For any $\alpha\in\mathbb{R}_+$ let $\underline{\underline{U}}(\alpha)$ be the set of open and closed sets U of X such that for any $\mu\in M$ with $\mathrm{Supp}(v\mu)\subset U$ we have $q_\xi(\varphi\mu) \leqslant \alpha q_\xi(\mu)$ and let $\mathbf{U}_\alpha := \overline{\bigcup_{U\in\underline{\underline{U}}(\alpha)} U}$. Let $\alpha\in\mathbb{R}_+$ Let $U,V\in\underline{\underline{U}}(\alpha)$. Then

$$U\cap V \ , \ U\setminus V \ , \ V\setminus U\in\underline{\underline{U}}(\alpha)$$

and by the above considerations

$$U\cup V = (U\cap V)\cup(U\setminus V)\cup(V\setminus U)\in\underline{\underline{U}}(\alpha) .$$

Hence $\underline{\underline{U}}(\alpha)$ is upper directed with respect to the inclusion relation. Let $\mu \in M$ with $\mathrm{Supp}\,(v\mu) \subset U_\alpha$. We have (Proposition 3.1.1 a), b), e))

$$\int |\xi| d|\varphi\mu| = \int \hat{u}|\xi| d(v|\varphi\mu|) = \sup_{U \in \underline{\underline{U}}(\alpha)} \int \hat{u}|\xi| d(1_U^Y \cdot v|\varphi\mu|) =$$

$$= \sup_{U \in \underline{\underline{U}}(\alpha)} \int |\xi| d((\hat{u}^{-1} 1_U^Y) \cdot |\varphi\mu|) = \sup_{U \in \underline{\underline{U}}(\alpha)} \int |\xi| d|\varphi((\hat{u}^{-1} 1_U^Y) \cdot \mu)| =$$

$$= \sup_{U \in \underline{\underline{U}}(\alpha)} q_\xi(\varphi((\hat{u}^{-1} 1_U^Y) \cdot \mu)) \leq \sup_{U \in \underline{\underline{U}}(\alpha)} \alpha q_\xi((\hat{u}^{-1} 1_U^Y) \cdot \mu) \leq \alpha \int |\xi| d|\mu|$$

and therefore $U_\alpha \in \underline{\underline{U}}(\alpha)$.

It is obvious that $U_\alpha \subset U_\beta$ for $\alpha, \beta \in \mathbb{R}_+$, $\alpha \leq \beta$. Let $x \in Y \setminus \bigcup_{\alpha \in \mathbb{R}_+} U_\alpha$. Then $x \in \mathrm{Supp}(\hat{u}\xi)$ and $\{x\}$ is not an isolated point of Y. Hence there exists a disjoint sequence $(K_n)_{n \in \mathbb{N}}$ of open, compact sets of Y whose union is relatively compact such that

$$K_n \cap U_n = \emptyset \ , \ 0 < \sup_{y \in K_n} |(\hat{u}\xi)(y)| \leq 2 \inf_{y \in K_n} |(\hat{u}\xi)(y)| =: \alpha_n$$

for any $n \in \mathbb{N}$. Let $(\mu_n)_{n \in \mathbb{N}}$ be a sequence in M_c such that

$$\mathrm{supp}(v\mu_n) \subset K_n \ , \ q_\xi(\varphi\mu_n) \geq n q_\xi(\mu_n)$$

for any $n \in \mathbb{N}$. Then (Proposition 3.1.1 a), b), c))

$$\beta_n := \int 1_X d|\mu_n| \leq \frac{2}{\alpha_n} \int |\xi| d|\mu_n| = \frac{2}{\alpha_n} q_\xi(\mu_n) \leq \frac{2}{n\alpha_n} q(\varphi\mu_n) =$$

$$= \frac{2}{n\alpha_n} \int |\xi| d(|\varphi\mu_n|) \leq \frac{2}{n} \int 1_X d(|\varphi\mu_n|)$$

for any $n \in \mathbb{N}$. We set

$$\mu := \sum_{n \in \mathbb{N}} \frac{1}{n^2 \beta_n} \mu_n \in M_c \ .$$

We deduce from the above inequalities

$$\int 1_X d|\varphi\mu| = \sum_{n\in\mathbb{N}} \frac{1}{n^2\beta_n} \int 1_X d|\varphi\mu_n| \geq \sum_{n\in\mathbb{N}} \frac{1}{2n} = \infty$$

and this as a contradiction. Hence $Y = \bigcup_{\alpha\in\mathbb{R}_+} U_\alpha$. By Proposition 1.6.2 and Proposition 3.1.1 c) there exists $\alpha\in\mathbb{R}_+$ with

$$\mathrm{Supp}\,(\hat{u}\xi) \subset U_\alpha$$

and we get by Proposition 3.1.1 a) $q_\xi \circ \varphi \leq \alpha q_\xi$.

c $\Rightarrow$ b follows from Theorem 3.2.2 a).

a $\Rightarrow$ c. By a $\Rightarrow$ b and Proposition 3.5.1 a) there exists a unique $\xi_\mu \in \hat{L}^1_{loc}(\mu)$ such that $\varphi\mu = \xi_\mu \cdot \mu$ for any $\mu\in M$.

Let $\mu\in M$ and assume $\xi_\mu \notin M^\pi_c$. Let (Y,u,v) be a representation of (X,M) (Theorem 2.3.8). By Proposition 3.1.1 a) c) there exists $x\in Y$ such that $|(\hat{u}\xi_\mu)(x)| = \infty$. Let $(K_n)_{n\in\mathbb{N}}$ be a sequence of compact open sets of Y such that

$$K_{n+1} \subset K_n \,, \quad \inf_{y\in K_n} |(\hat{u}\xi_\mu)(y)| \geq n$$

for any $n\in\mathbb{N}$. We set $\eta_n := \hat{u}^{-1} 1^Y_{K_n} \in M^\pi$ (Proposition 3.1.1 b) c)) for any $n\in\mathbb{N}$. By Theorem 3.1.7 b) $\eta_n^2 = \eta_n$ and by Proposition 3.5.1 d) $\xi_{\eta_n\cdot\mu} = \eta_n \xi_\mu$ for any $n\in\mathbb{N}$. Let $\alpha\in\mathbb{R}_+$ such that $q_{\eta_1} \circ \varphi \leq \alpha q_{\eta_1}$. We get (Theorem 3.4.15 e), Theorem 3.2.2 d), Proposition 3.1.1 a) b), Theorem 3.1.7)

$$n q_{\eta_1}(\eta_n\cdot\mu) = n\int \eta_1 d|\eta_n\cdot\mu| = n\int \eta_1\eta_n d|\mu| = n\int 1^Y_{K_1} 1^Y_{K_n} d|v\mu| \leq$$

$$\leq \int 1^Y_{K_1} 1^Y_{K_n} |\hat{u}\xi_\mu| d|v\mu| = \int \eta_1\eta_n |\xi_\mu| d|\mu| = \int \eta_1 d|(\eta_n\xi_\mu)\cdot\mu| =$$

$$= q_{\eta_1}(\varphi(\eta_n\cdot\mu)) \leq \alpha q_{\eta_1}(\eta_n\cdot\mu)$$

for any $n\in\mathbb{N}$. We get

$$\int 1_{K_n}^{Y} d(\nu\mu) = 0$$

for $n > \alpha$ and this is a contradiction. Hence $\xi_\mu \in M_c^\pi$.

Let $\mu \in M$ and let F be the set of step functions on X with respect to $\underline{\underline{R}}$. By Proposition 3.5.1 b) c) d)

$$\varphi(f\cdot\mu) = \xi_\mu\cdot(f\cdot\mu)$$

for any $f \in F$. By c $\Longrightarrow$ a and Proposition 3.4.3 $\varphi\nu = \xi_\mu\cdot\nu$ for any ν belonging to the band of M generated by μ .

Let (Y,u,v) be a representation of (X,M) (Theorem 2.3.8). By the above considerations $\hat{u}\xi_\mu = \hat{u}\xi_\nu$ on Supp $(v\mu) \cap$ Supp $(v\nu)$ for any $\mu,\nu \in M$. By Proposition 2.4.4 a) there exists a unique $g \in C_\infty(Y)$ such that $g = \hat{u}\xi_\mu$ on Supp $(v\mu)$ for any $\mu \in M$. Let $x \in Y$ such that $|g(x)| = \infty$. Then there exists a sequence $(K_n)_{n\in\mathbb{N}}$ of compact sets of Y whose union is relatively compact and such that $|g| \geqslant n$ on K_n for any $n \in \mathbb{N}$. Let $(\mu_n)_{n\in\mathbb{N}}$ be a sequence in M_+ with

$$\text{supp}\,(v\mu_n) \subset K_n \quad , \quad \int 1_{K_n}^{Y} d(v\mu_n) = 1$$

for any $n \in \mathbb{N}$. Then

$$\mu := \sum_{n\in\mathbb{N}} \frac{1}{n^2}\mu_n \in M$$

and g is not bounded on Supp $(v\mu)$ and this is a contradiction. Hence $g \in C(Y)$. We set $\eta := \hat{u}^{-1} g \in M_c^\pi$ (Proposition 3.1.1 b) c)). It is easy to see that $\varphi\mu = \eta\mu$ for any $\mu \in M$. □

<u>Remark</u>. The hypothesis $\varphi\mu \ll \mu$ for any $\mu \in M_b$ is not equivalent to the above assertions even if φ is positive. Indeed let $\underline{\underline{F}}$ be a non-trivial ultrafilter on X such that the intersection of any sequence in $\underline{\underline{F}}$ belongs to $\underline{\underline{F}}$, let $\underline{\underline{R}}$ be the set of finite subsets of X, let M be $M(\underline{\underline{R}})$, and let $\mu_o \in M$, $\mu_o \neq 0$. The map

$$\varphi : M \longrightarrow M \ , \ \mu \longmapsto (\lim_{x,\underline{\underline{F}}} \mu(\{x\}))\cdot\mu_o$$

does not belong to $L_d(M)$ but we have $\varphi\mu \ll \mu$ for any $\mu \in M_b$.

Theorem 3.5.4 *For any* $\varphi \in L_d(M)$ *there exists a unique* $\tilde{\varphi} \in M_c^{\pi}$ *such that* $\varphi\mu = \tilde{\varphi}\cdot\mu$ *for any* $\mu \in M$ *. The map*

$$L_d(M) \longrightarrow M_c^{\pi}, \quad \varphi \longmapsto \tilde{\varphi}$$

is an isomorphism of unital algebras and an isomorphism of ordered sets such that $r_{\xi}(\varphi) = q_{\xi}(\tilde{\varphi})$ *for any* $(\varphi,\xi) \in L_d(M) \times M^{\pi}$ *.*

Let $\varphi \in L_d(M)$. By Theorem 3.5.3 (a $\Longrightarrow$ c) there exists an element $\tilde{\varphi} \in M_c^{\pi}$ such that $\varphi\mu = \tilde{\varphi}\cdot\mu$ for any $\mu \in M$. The unicity of $\tilde{\varphi}$ follows from Theorem 3.4.15 f). Let $\varphi, \psi \in L_d(M)$ and let $\alpha, \beta \in \mathbb{R}$. Then (Theorem 3.4.15)

$$\widetilde{(\alpha\varphi+\beta\psi)}\cdot\mu = (\alpha\varphi+\beta\psi)(\mu) = \alpha\varphi\mu+\beta\psi\mu = \alpha\tilde{\varphi}\cdot\mu+\beta\tilde{\psi}\cdot\mu = (\alpha\tilde{\varphi}+\beta\tilde{\psi})\cdot\mu ,$$

$$(\widetilde{\varphi\circ\psi})\cdot\mu = \varphi(\psi\mu) = \varphi(\tilde{\psi}\cdot\mu) = (\tilde{\varphi}\tilde{\psi})\cdot\mu ,$$

$$\varphi\mu \leq \psi\mu \Longleftrightarrow (\tilde{\psi}-\tilde{\varphi})\cdot\mu \geq 0$$

for any $\mu \in M_+$ and therefore (Theorem 3.4.15 f))

$$\widetilde{\alpha\varphi+\beta\psi} = \alpha\tilde{\varphi}+\beta\tilde{\psi} , \quad \widetilde{\varphi\circ\psi} = \tilde{\varphi}\tilde{\psi} , \quad \varphi \leq \psi \Longleftrightarrow \tilde{\varphi} \leq \tilde{\psi} .$$

Hence the map

$$L_d(M) \longrightarrow M_c^{\pi}, \quad \varphi \longrightarrow \tilde{\varphi}$$

is an isomorphism of unital algebras and an isomorphism of ordered sets (Theorem 3.5.3 (c $\Longrightarrow$ a)).

The last assertion is obvious. □

Theorem 3.5.5 *Let* φ *be a linear map* $M^{\pi} \longrightarrow M^{\pi}$ *. The following assertions are equivalent:*

a) $\xi\varphi\eta = 0$ *for any* $\xi, \eta \in M^{\pi}$ *with* $\xi\eta = 0$ *;*

b) there exists $\zeta \in M_c^{\pi}$ *with* $\varphi\xi = \zeta\xi$ *for any* $\xi \in M^{\pi}$ *.*

Let φ' be the adjoint map $M \longrightarrow M$ of φ (Theorem 3.3.1).
a $\longrightarrow$ b. Let $\mu \in M$ and let $A \in \underline{\underline{R}}$ be a μ-null set. Then $i_A \in M^{\pi} \cap \hat{L}(\mu)$

and we get

$$\xi\varphi i_A = 0$$

for any $\xi \in M^\pi \cap \hat{L}^\perp(\mu)$. Hence $\varphi i_A \in \hat{L}(\mu)$ and therefore

$$\varphi'\mu(A) = \langle\varphi'\mu, i_A\rangle = \langle\mu, \varphi\, i_A\rangle = 0 \quad .$$

Since A is arbitrary we deduce $\varphi'\mu \ll \mu$. By Theorem 3.5.3 (b $\Longrightarrow$ c) there exists $\zeta \in M_c^\pi$ such that $\varphi'\mu = \zeta\cdot\mu$ for any $\mu \in M$. Let $\xi \in M^\pi$. We get (Theorem 3.4.15)

$$\langle\mu,\varphi\xi\rangle = \langle\varphi'\mu,\xi\rangle = \langle\zeta\cdot\mu,\xi\rangle = \langle\mu,\zeta\xi\rangle$$

for any $\mu \in M$ and therefore $\varphi\xi = \zeta\xi$.

b $\Longrightarrow$ a is trivial. □

6. The spaces M_b and M_b^π

Definition 3.6.1 *For any* $\xi \in M_b^\pi$ *and for any* $\mu \in M_b$ *we set*

$$\|\xi\| := p_{i_X}(\xi) = \inf\{\alpha \in \mathbb{R}_+ \mid |\xi| \leq \alpha i_X\},\quad \|\mu\| := q_{i_X}(\mu) = \int 1_X d|\mu| \quad .$$

Theorem 3.6.2

a) The map

$$M_b^\pi \to \mathbb{R}\ ,\ \xi \longmapsto \|\xi\|$$

is an M-norm such that

$$\|\xi\eta\| \leq \|\xi\|\ \|\eta\|$$

for any $\xi,\eta \in M_b^\pi$; M_b^π *endowed with it is an order complete Banach lattice and a Banach algebra;*

b) the map

$$M_b \to \mathbb{R}\ ,\ \mu \to \|\mu\|$$

is an L-norm; M_b *endowed with it is an order complete Banach lattice whose dual is* M_b^π *;*

c) *we have* $\|\xi\mu\| \leq \|\xi\| \, \|\mu\|$ *for any* $(\xi,\mu) \in M_b^\pi \times M_b$ *;* M_b *endowed with the action*

$$M_b^\pi \times M_b \longrightarrow M_b \ , \ (\xi,\mu) \longmapsto \xi\cdot\mu$$

is a Banach M_b^π*-module with respect to the above norms.*

The assertions follow e.g. from Theorem 2.3.8 Proposition 3.1.1 b), c), e) and Theorem 3.1.7 g). □

<u>*Theorem 3.6.3*</u> *Let* $\mu \in M$ *and let* $S(\mu)$ *be the solid subspace of* M *generated by* μ *. We set*

$$\hat{L}^\infty(\mu) := M_b^\pi \cap \hat{L}^\perp(\mu)$$

and endow $\hat{L}^\infty(\mu)$ *with the structures induced by* M_b^π *inclusively the norm defined in Theorem 3.6.2 a).*

a) $\hat{L}^\infty(\mu)$ *is an order complete Banach lattice and a Banach algebra;*

b) *for any* $\xi \in \hat{L}^\infty(\mu)$ *we have* $\xi\cdot\mu \in S(\mu)$ *and the map*

$$\hat{L}^\infty(\mu) \longrightarrow S(\mu) \ , \ \xi \longmapsto \xi\cdot\mu$$

is a linear bijection; if μ *is positive it is an isomorphism of vector lattices;*

c) *for any* $\xi \in \hat{L}^\infty(\mu)$ *the map*

$$\xi' : L^1(\mu) \longrightarrow \mathbb{R} \ , \ f \longmapsto \int f d(\xi\cdot\mu)$$

belongs to the dual of $L^1(\mu)$ *and the map*

$$\hat{L}^\infty(\mu) \longrightarrow (L^1(\mu))' \ , \ \xi \longmapsto \xi'$$

is an isomorphism of Banach spaces; if μ *is positive it is an isomorphism of Banach lattices;*

d) *for any* $\xi \in \hat{L}^1(\mu)$ *the map*

$$\xi' : \hat{L}^{\infty}(\mu) \longmapsto \mathbb{R} , \eta \longmapsto \int \xi \eta d\mu$$

belongs to $\hat{L}^{\infty}(\mu)^{\pi}$ *and the map*

$$\hat{L}^{1}(\mu) \longrightarrow \hat{L}^{\infty}(\mu)^{\pi} , \xi \longrightarrow \xi'$$

is an isomorphism of Banach spaces; it is an isomorphism of Banach lattices if μ *is positive.*

The assertions follow immediately from Theorem 2.1.3, Theorem 2.3.7 b), Proposition 3.1.1 a), b), c), e), Theorem 3.1.7 g), Theorem 3.6.2, and Theorem 3.2.4 f). □

Theorem 3.6.4 *Let* $\mu \in M_{+}$. *Then:*

a) *for any* $(\xi,\eta) \in \hat{L}^{\infty}(\mu) \times \hat{L}^{1}(\mu)$ *we have* $\xi\eta \in \hat{L}^{1}(\mu)$ *and*

$$\int |\xi\eta| d\mu \leq \|\xi\| \int |\eta| d\mu \ ;$$

b) $\hat{L}^{1}(\mu)$ *endowed with the action*

$$\hat{L}^{\infty}(\mu) \times \hat{L}^{1}(\mu) \longrightarrow \hat{L}^{1}(\mu) , (\xi,\eta) \longmapsto \xi\eta$$

is a Banach $\hat{L}^{\infty}(\mu)$*-module.* □

7. Tensor products of spaces of measures

Throughout this section we denote by X_1, X_2 *two sets, by* $\underline{\underline{R}}_1$, $\underline{\underline{R}}_2$ δ*-rings of subsets of* X_1 *and* X_2 *respectively, by* M_1 , M_2 *bands of the vector lattices of real measures on* $\underline{\underline{R}}_1$ *and* $\underline{\underline{R}}_2$ *respectively, by* $\underline{\underline{R}}_1 \otimes \underline{\underline{R}}_2$ *and* $\underline{\underline{S}}$ *the* δ*-ring and the ring of sets generated by* $\{A_1 \times A_2 \mid (A_1, A_2) \in \underline{\underline{R}}_1 \times \underline{\underline{R}}_2\}$ *respectively, and for any* $(\mu_1,\mu_2) \in M_1 \times M_2$ *by* $\mu_1 \tilde{\times} \mu_2$ *the product of these measures (i.e. the measure on* $\underline{\underline{R}}_1 \otimes \underline{\underline{R}}_2$ *equal to* $\mu_1(A_1)\mu_2(A_2)$ *at* $A_1 \times A_2$ *for any* $(A_1,A_2) \in \underline{\underline{R}}_1 \times \underline{\underline{R}}_2$). *For any sets* Y, Z *with* $Z \subset Y$ *we denote by* 1_Z^Y *the real function on* Y *equal to* 1 *on* Z *and equal to* 0 *on* $Y \setminus Z$. *The notations and the terminology introduced in the preceding sections with respect to* $(X, \underline{\underline{R}}, M)$ *will be used with respect to* $(X_1, \underline{\underline{R}}_1, M_1)$, $(X_2, \underline{\underline{R}}_2, M_2)$ *in a similar way, the notations wearing the sufixes* 1 *and* 2 *respectively.*

We assume:

a) $X = X_1 \times X_2$;

b) $\underline{\underline{R}} = \underline{\underline{R}}_1 \otimes \underline{\underline{R}}_2$;

c) M *is the band of* $M(\underline{\underline{R}})$ *generated by* $\{\mu_1 \tilde{\times} \mu_2 \mid (\mu_1, \mu_2) \in M_1 \times M_2\}$.

Let N_1, N_2 be subspaces of M_1 and M_2 respectively. The map

$$N_1 \times N_2 \longrightarrow M \, , \; (\mu_1, \mu_2) \longmapsto \mu_1 \tilde{\times} \mu_2$$

being bilinear it generates a canonical linear map $N_1 \otimes N_2 \longrightarrow M$; this map is injective. *We identify* $N_1 \otimes N_2$ *with its image in* M *via this injective map and denote by* $N_1 \bar{\otimes} N_2$ *the solid subspace of* M *generated by* $N_1 \otimes N_2$. If N_1, N_2 are fundamental solid subspaces of M_1 and M_2 respectively then $N_1 \bar{\otimes} N_2$ is a fundamental solid subspace of M . We have $|\mu_1 \otimes \mu_2| = |\mu_1| \otimes |\mu_2|$ for any $\mu_1, \mu_2 \in M_1 \times M_2$.

Let N_1, N_2 be fundamental solid subspaces of M_1 and M_2 respectively and let F_1, F_2 be subsets of N_1^π and N_2^π respectively. *We denote by* $(N_1, F_1) \otimes (N_2, F_2)$ *the vector space* $N_1 \otimes N_2$ *endowed with the topology of bi-equicontinuous convergence ([26] page 96) and by* $(N_1, F_1) \tilde{\otimes} (N_2, F_2)$ *its completion.*

For any real functions f_1, f_2 *on* X_1 *and* X_2 *respectively we set*

$$f_1 \tilde{\times} f_2 : X_1 \times X_2 \longrightarrow \mathbb{R} \, , \; (x_1, x_2) \longmapsto f_1(x_1)\, f_2(x_2) .$$

Proposition 3.7.1 *Let* N_1, N_2 *be solid subspaces of* M_1 *and* M_2 *respectively, let* P *be the set*

$$\{ \sum_{\iota \in I} \mu_{1\iota} \otimes \mu_{2\iota} \mid (\mu_{1\iota}, \mu_{2\iota})_{\iota \in I} \text{ finite family in } N_{1+} \times N_{2+} \}$$

and let f *be a step function on* X *with respect to* $\underline{\underline{S}}$. *Then*

a) $f \geq 0, \; \mu \in P \Longrightarrow f \cdot \mu \in P$;

b) $\mu \in N_1 \otimes N_2 \Longrightarrow f \cdot \mu \in N_1 \otimes N_2$,

a) There exist finite families $(\mu_{1\iota}, \mu_{2\iota})_{\iota \in I}$, $(A_{1\kappa}, A_{2\kappa})_{\kappa \in K}$, $(\alpha_\kappa)_{\kappa \in K}$ in $N_{1+} \times N_{2+}$, $\underline{\underline{R}}_1 \times \underline{\underline{R}}_2$, and $\mathbb{R}_+$ respectively such that

$$\mu = \sum_{\iota\in I} \mu_{1\iota} \otimes \mu_{2\iota}, \quad f = \sum_{\kappa\in K} \alpha_\kappa 1^X_{A_{1\kappa}\times A_{2\kappa}}$$

and we get

$$f\cdot\mu = \sum_{(\iota,\kappa)\in I\times K} (\alpha_\kappa 1^{X_1}_{A_{1\kappa}} \cdot \mu_{1\iota}) \otimes (1^{X_2}_{A_{2\kappa}} \cdot \mu_{2\iota}) \in P .$$

b) follows immediately from a). □

<u>*Proposition 3.7.2*</u> *Let* N_1, N_2 *be solid subspaces of* M_1 *and* M_2 *respectively, let* P *be the set*

$$\{ \sum_{\iota\in I} \mu_{1\iota}\otimes\mu_{2\iota} \mid (\mu_{1\iota},\mu_{2\iota})_{\iota\in I} \text{ finite family in } N_{1+}\times N_{2+}\} ,$$

let N *be a fundamental solid subspace of* M *containing* $N_1 \otimes N_2$, *let* B *be the band of* N *generated by* $N_1 \otimes N_2$, *and let* F *be a subset of* N^π *generating* N^π *as band. Then* B *and* B_+ *are the closures in* (N,F) *of* $N_1 \otimes N_2$ *and* P *respectively.*

The assertion follows immediately from Proposition 3.4.7 and 3.7.1. □

<u>*Corollary 3.7.3*</u> *Let* N_1, N_2 *be fundamental solid subspaces of* M_1 *and* M_2 *respectively and let* F *be a subset of* M^π *generating it as band. Then* $N_1 \times N_2$ *is dense in* (M,F) *and* M_+ *is the closure of*

$$\{ \sum_{\iota\in I} \mu_{1\iota}\otimes\mu_{2\iota} \mid (\mu_{1\iota},\mu_{2\iota})_{\iota\in I} \text{ finite family in } N_{1+} \times N_{2+}\}$$

in (M,F) . □

<u>*Corollary 3.7.4*</u> *Let* N_1, N_2 *be fundamental solid subspaces of* M_1 *and* M_2 *respectively and let* $\xi\in(N_1 \bar{\otimes} N_2)^\pi$. *If* $\int\xi d(\mu_1\otimes\mu_2) \geq 0$ *for any* $(\mu_1,\mu_2)\ N_{1+} \times N_{2+}$ *then* ξ *is positive. If* $\int\xi d(\mu_1\otimes\mu_2) = 0$ *for any* $(\mu_1,\mu_2)\in N_{1+} \times N_{2+}$ *then* $\xi = 0$. □

<u>*Proposition 3.7.5*</u> *If* N_1, N_2 *are solid subspaces of* M_1 *and* M_2 *respectively then*

$$N_1 \bar{\otimes} N_2 = \{\mu\in M \mid \exists(\mu_1,\mu_2)\in N_1 \times N_2,\ |\mu| \leq |\mu_1 \otimes \mu_2|\}$$

Let $\mu \in N_1 \otimes N_2$. Then there exists a finite family $(\mu_{1\iota}, \mu_{2\iota})_{\iota \in I}$ in $N_1 \times N_2$ such that

$$|\mu| \leqslant \sum_{\iota \in I} |\mu_{1\iota} \otimes \mu_{2\iota}| .$$

If we set

$$\mu_1 := \sum_{\iota \in I} |\mu_{1\iota}| \in N_1 , \quad \mu_2 := \sum_{\iota \in I} |\mu_{2\iota}| \in N_2$$

then

$$|\mu| \leqslant \mu_1 \otimes \mu_2 .$$

The other inclusion is trivial. □

Theorem 3.7.6 *Let* $(\xi_1, \xi_2) \in M_1^\rho \times M_2^\rho$. *There exists a unique element of* M^ρ, *which will be denoted by* $\xi_1 \tilde{\times} \xi_2$, *such that for any* $(\mu_1, \mu_2) \in M_{1c+} \times M_{2c+}$ *with* $(\xi_1, \xi_2) \in \hat{\mathcal{L}}^1(\mu_1) \times \hat{\mathcal{L}}^1(\mu_2)$ *we have*

$$\xi_1 \tilde{\times} \xi_2 \in \hat{\mathcal{L}}^1(\mu_1 \otimes \mu_2), \int \xi_1 \tilde{\times} \xi_2 d(\mu_1 \otimes \mu_2) = (\int \xi_1 d\mu_1)(\int \xi_2 d\mu_2) .$$

We have:

a) $|\xi_1 \tilde{\times} \xi_2| = |\xi_1| \tilde{\times} |\xi_2|$;

b) the following assertions hold for any $(\mu_1, \mu_2) \in M_1 \times M_2$:

b_1) $(\xi_1, \xi_2) \in \hat{\mathcal{L}}^1(\mu_1) \times \hat{\mathcal{L}}^1(\mu_2) \Longrightarrow \xi_1 \tilde{\times} \xi_2 \in \hat{\mathcal{L}}^1(\mu_1 \otimes \mu_2)$,
$\int \xi_1 \tilde{\times} \xi_2 d(\mu_1 \otimes \mu_2) = (\int \xi_1 d\mu_1)(\int \xi_2 d\mu_2)$;

b_2) $(\xi_1, \xi_2) \in \hat{\mathcal{L}}^1_{loc}(\mu_1) \times \hat{\mathcal{L}}^1_{loc}(\mu_2) \Longrightarrow \xi_1 \tilde{\times} \xi_2 \in \hat{\mathcal{L}}^1_{loc}(\mu_1 \otimes \mu_2)$,
$(\xi_1 \tilde{\times} \xi_2) \cdot (\mu_1 \otimes \mu_2) = (\xi_1 \cdot \mu_2) \otimes (\xi_2 \cdot \mu_2)$.

c) if there exists $(f_1, f_2) \in L_{1\infty} \times L_{2\infty}$ *with* $\dot{f}_1 = \xi_1$, $\dot{f}_2 = \xi_2$ *then*

$$\xi_1 \tilde{\times} \xi_2 = \dot{\widehat{f_1 \tilde{\times} f_2}} .$$

$M_1(\xi_1)$, $M_2(\xi_2)$ are fundamental solid subspaces of M_1 and M_2 respectively and therefore $M_1(\xi_1) \bar{\otimes} M_2(\xi_2)$ is a fundamental solid subspace of M. Let $\mu \in M_1(\xi_1) \bar{\otimes} M_2(\xi_2)$. By Proposition 3.7.5 there exists $(\mu_1,\mu_2) \in M_1(\xi_1) \times M_2(\xi_2)$ such that $|\mu| \leqslant |\mu_1 \otimes \mu_2|$. By Theorem 3.2.2 d) and Propositions 1.4.1 and 1.4.4 there exists $(f_1,f_2) \in L^1(\mu_1) \times L^1(\mu_2)$ such that $\xi_1 \cdot \mu_1 = f_1 \cdot \mu_1$, $\xi_2 \cdot \mu_2 = f_2 \cdot \mu_2$. Then $f_1 \tilde{\times} f_2 \in L^1(\mu)$ and

$$\int f_1 \tilde{\times} f_2 d\mu$$

does depend neither on the choise of f_1, f_2 nor on the choice of μ_1, μ_2; hence we may set

$$\mu^* := \int f_1 \tilde{\times} f_2 d\mu .$$

We denote by $\xi_1 \tilde{\times} \xi_2$ the map

$$M_1(\xi_1) \bar{\otimes} M_2(\xi_2) \longrightarrow \mathbb{R} , \mu \longmapsto \mu^* .$$

It is obvious that $\xi_1 \tilde{\times} \xi_2$ belongs to $(M_1(\xi_1) \bar{\otimes} M_2(\xi_2))^\pi$ and therefore to M^ρ and that b_1) and c) are fulfilled.

The uniqueness as well as a) follow immediately from the above considerations and Corollary 3.7.4.

In order to prove b_2) let $(\xi_1,\xi_2) \in \hat{L}^1_{loc}(\mu_1) \times \hat{L}^1_{loc}(\mu_2)$. We have for any $(A_1,A_2) \in \underline{\underline{R}}_1 \times \underline{\underline{R}}_2$

$$(\xi_1,\xi_2) \in \hat{L}^1(1^{X_1}_{A_1} \cdot \mu_1) \times \hat{L}^1(1^{X_2}_{A_2} \cdot \mu_2)$$

and by a) and Theorem 3.2.2 d)

$$\xi_1 \tilde{\times} \xi_2 \in \hat{L}^1((1^{X_1}_{A_1} \cdot \mu_1) \otimes (1^{X_2}_{A_2} \cdot \mu_2)) = L^1(1^{X}_{A_1 \times A_2} \cdot (\mu_1 \otimes \mu_2)),$$

$$\int \xi_1 \tilde{\times} \xi_2 d(1^{X}_{A_1 \times A_2} \cdot (\mu_1 \otimes \mu_2)) = \int \xi_1 \tilde{\times} \xi_2 d((1^{X_1}_{A_1} \cdot \mu_1) \otimes (1^{X_2}_{A_2} \cdot \mu_2)) =$$

$$= (\int \xi_1 d(1^{X_1}_{A_1} \cdot \mu_1))(\int \xi_2 d(1^{X_2}_{A_2} \cdot \mu_2) = ((\xi_1 \cdot \mu_1)(A_1))((\xi_2 \cdot \mu_2)(A_2)) =$$

$$= ((\xi_1 \cdot \mu_1) \otimes (\xi_2 \cdot \mu_2))(A_1 \times A_2) .$$

Hence $\xi_1 \tilde{\times} \xi_2 \in \hat{\mathcal{L}}^1_{loc}(\mu_1 \otimes \mu_2)$ and $(\xi_1 \tilde{\times} \xi_2) \cdot (\mu_1 \otimes \mu_2) = (\xi_1 \cdot \mu_1) \otimes (\xi_2 \cdot \mu_2)$. □

Proposition 3.7.7

a) The map

$$M_1^\rho \times M_2^\rho \longrightarrow M^\rho, \ (\xi_1, \xi_2) \longmapsto \xi_1 \tilde{\times} \xi_2$$

is bilinear;

b) the linear map $M_1^\rho \otimes M_2^\rho \longrightarrow M^\rho$ generated by the above bilinear map is injective and a homomorphism of unital algebras.

a) Let $(\xi_1', \xi_1'', \xi_2) \in M_1^\rho \times M_1^\rho \times M_2^\rho$ and let α' , $\alpha'' \in \mathbb{R}$. Let $(\mu_1, \mu_2) \in M_{1\mathbb{C}} \times M_{2\mathbb{C}}$ such that

$$(\xi_1', \xi_1'', \xi_2) \in \hat{\mathcal{L}}^1(\mu_1) \times \hat{\mathcal{L}}^1(\mu_1) \times \hat{\mathcal{L}}^1(\mu_2) .$$

By Corollary 3.1.3 a) we get

$$(\alpha'\xi_1' + \alpha''\xi_1'', \ \xi_2) \in \hat{\mathcal{L}}^1(\mu_1) \times \hat{\mathcal{L}}^1(\mu_2) ,$$

$$\int (\alpha'\xi_1' + \alpha''\xi_1'') \tilde{\times} \xi_2 d(\mu_1 \otimes \mu_2) = (\int (\alpha'\xi_1' + \alpha''\xi_1'') d\mu_1)(\int \xi_2 d\mu_2) =$$

$$= \alpha'(\int \xi_1' d\mu_1)(\int \xi_2 d\mu_2) + \alpha''(\int \xi_1'' d\mu_1)(\int \xi_2 d\mu_2) =$$

$$= \alpha' \int \xi_1' \tilde{\times} \xi_2 \, d(\mu_1 \otimes \mu_2) + \alpha'' \int \xi_1'' \tilde{\times} \xi_2 d(\mu_1 \otimes \mu_2) =$$

$$= \int (\alpha'(\xi_1' \tilde{\times} \xi_2) + \alpha''(\xi_1'' \tilde{\times} \xi_2)) d(\mu_1 \otimes \mu_2) .$$

By Corollary 3.7.4 we get

$$(\alpha'\xi_1' + \alpha''\xi_1'') \tilde{\times} \xi_2 = \alpha'(\xi_1' \tilde{\times} \xi_2) + \alpha''(\xi_1'' \tilde{\times} \xi_2) .$$

Hence the map

$$M_1^\rho \times M_2^\rho \longrightarrow M^\rho, \ (\xi_1, \xi_2) \longmapsto \xi_1 \tilde{\times} \xi_2$$

is bilinear.

b) It is obvious that the map $M_1^\rho \times M_2^\rho \longrightarrow M^\rho$ is injective and $i_{X_1}^{X_1} \tilde{\times} i_{X_2}^{X_2} = i_X^X$. Let (ξ_1,ξ_2), $(\eta_1,\eta_2) \in M_1^\rho \times M_2^\rho$ and let $(\mu_1,\mu_2) \in M_1 \times M_2$ such that

$$\xi_1, \eta_1, \xi_1\eta_1 \in \hat{\mathcal{L}}^1(\mu_1), \quad \xi_2, \eta_2, \xi_2\eta_2 \in \hat{\mathcal{L}}^1(\mu_2).$$

By Theorems 3.7.6 b) and 3.2.2 d) we have

$$\int (\xi_1\eta_1) \tilde{\times} (\xi_2\eta_2) d(\mu_1 \otimes \mu_2) = (\int \xi_1\eta_1 d\mu_1)(\int \xi_2\eta_2 d\mu_2) =$$

$$= (\int \xi_1 d(\eta_1 \cdot \mu_1))(\int \xi_2 d(\xi_2 \cdot \mu_2)) = \int (\xi_1 \tilde{\times} \xi_2) d((\eta_1 \cdot \mu_1) \otimes (\eta_2 \mu_2)) =$$

$$= \int (\xi_1 \tilde{\times} \xi_2) d((\eta_1 \tilde{\times} \eta_2) \cdot (\mu_1 \otimes \mu_2)) = \int (\xi_1 \tilde{\times} \xi_2)(\eta_1 \tilde{\times} \eta_2) d(\mu_1 \otimes \mu_2).$$

By Corollary 3.7.4

$$(\xi_1 \tilde{\times} \xi_2)(\eta_1 \tilde{\times} \eta_2) = (\xi_1\eta_1) \tilde{\times} (\xi_2\eta_2).$$

Hence the map $M_1^\rho \otimes M_2^\rho \longrightarrow M^\rho$ is a homomorphism of unital algebras. □

We identify $M_1^\rho \otimes M_2^\rho$ *with a unital subalgebra of* M^ρ *via the above injection* $M_1^\rho \otimes M_2^\rho \longrightarrow M^\rho$.

Proposition 3.7.8

a) For any upper bounded family $(\xi_{1\iota})_{\iota \in I}$ *in* M_1^ρ *and for any* $\xi_2 \in M_{2+}^\rho$ *we have*

$$(\bigvee_{\iota \in I} \xi_{1\iota}) \otimes \xi_2 = \bigvee_{\iota \in I} (\xi_{1\iota} \otimes \xi_2);$$

b) if F_1, F_2 *are subsets of* M_1^ρ *and* M_2^ρ *respectively generating the corresponding spaces as bands then*

$$\{\xi_1 \otimes \xi_2 \mid (\xi_1,\xi_2) \in F_1 \times F_2\}$$

generates M^ρ *as band;*

c) we have

$$\xi_1 \otimes \xi_2 \in \hat{\mathcal{L}}^1(\mu_1 \otimes \mu_2) ,$$

$$\int |\xi_1 \otimes \xi_2| d|\mu_1 \otimes \mu_2| \neq 0 \implies (\xi_1, \xi_2) \in \hat{\mathcal{L}}^1(\mu_1) \times \hat{\mathcal{L}}^1(\mu_2)$$

for any $(\xi_1, \xi_2) \in M_1^\rho \times M_2^\rho$ *and for any* $(\mu_1, \mu_2) \in M_1 \times M_2$;

d) if N_1 , N_2 *are fundamental solid subspaces of* M_1 *and* M_2 *respectively then* $\xi_1 \otimes \xi_2 \in (N_1 \bar{\otimes} N_2)^\pi$ *for any* $(\xi_1, \xi_2) \in N_1^\pi \times N_2^\pi$.

a) By Proposition 3.7.7 a) and Theorem 3.7.6 a) the map

$$M_1^\rho \longrightarrow M^\rho , \ \xi_1 \longmapsto \xi_1 \otimes \xi_2$$

is a homomorphism of vector lattices and therefore we may assume the family $(\xi_{1\iota})_{\iota \in I}$ upper directed. We get for any $(\mu_1, \mu_2) \in M_{1c+} \times M_{2c+}$,

$$\int (\bigvee_{\iota \in I} \xi_{1\iota}) \otimes \xi_2 d(\mu_1 \otimes \mu_2) = (\int \bigvee_{\iota \in I} \xi_{1\iota} d\mu_1)(\int \xi_2 d\mu_2) =$$

$$= \sup_{\iota \in I} \int \xi_{1\iota} d\mu_1)(\int \xi_2 d\mu_2) = \sup_{\iota \in I} (\int \xi_{1\iota} d\mu_1)(\int \xi_2 d\mu_2) =$$

$$= \sup_{\iota \in I} (\int \xi_{1\iota} \otimes \xi_2 d(\mu_1 \otimes \mu_2)) = \int \bigvee_{\iota \in I} (\xi_{1\iota} \otimes \xi_2) d(\mu_1 \otimes \mu_2)$$

and therefore

$$(\bigvee_{\iota \in I} \xi_{1\iota}) \otimes \xi_2 = \bigvee_{\iota \in I} (\xi_{1\iota} \otimes \xi_2) .$$

b) Let B be the band of M^ρ generated by $\{\xi_1 \otimes \xi_2 \mid (\xi_1, \xi_2) \in F_1 \times F_2\}$ By Theorem 3.7.6 a) we have

$$\{|\xi_1| \otimes |\xi_2| \mid (\xi_1, \xi_2) \in F_1 \times F_2\} \subset B .$$

By a) we get successively

$$\{|\xi_1| \otimes \xi_2 \mid (\xi_1, \xi_2) \in F_1 \times F_2\} \subset B ,$$

$$\{|\xi_1| \times i_{A_2}^{X_2} \mid (\xi_1, A_2) \in F_1 \times \underline{\underline{R}}_2\} \subset B,$$

$$\{\xi_1 \otimes i^{X_2}_{A_2} \mid (\xi_1, A_2) \in F_1 \times \underline{\underline{R}}_2\} \subset \mathcal{B} ,$$

$$\{i^{X_1}_{A_1} \otimes i^{X_2}_{A_2} \mid (A_1, A_2) \in \underline{\underline{R}}_1 \times \underline{\underline{R}}_2\} \subset \mathcal{B} .$$

Let ξ be an element of M^ρ such that $|\xi| \wedge |\eta| = 0$ for any $\eta \in \mathcal{B}$. By the above relation we get

$$|\xi| \wedge (i^{X_1}_{A_1} \tilde{\otimes} i^{X_2}_{A_2}) = 0$$

for any $(A_1, A_2) \in \underline{\underline{R}}_1 \times \underline{\underline{R}}_2$. By Theorem 3.7.6 c) we get further

$$|\xi| \wedge i^X_{A_1 \times A_2} = 0$$

for any $(A_1, A_2) \in \underline{\underline{R}}_1 \times \underline{\underline{R}}_2$ which implies $\xi = 0$. Hence $\mathcal{B} = M^\rho$.

c) By a), Proposition 3.7.7 a) and Theorem 3.7.6 a) b) we have for any $(A_1, A_2, n) \in \underline{\underline{R}}_1 \times \underline{\underline{R}}_2 \times \mathbb{N}$

$$(|\xi_1| \wedge n i^{X_1}_{A_1}) \otimes (|\xi_2| \wedge n i^{X_2}_{A_2}) \in \hat{\mathcal{L}}^1(\mu_1 \otimes \mu_2) ,$$

$$(\int |\xi_1| \wedge n i^{X_1}_{A_1} d|\mu_1|)(\int |\xi_2| \wedge n i^{X_2}_{A_2} d|\mu_2|) =$$

$$= \int (|\xi_1| \wedge n i^{X_1}_{A_1}) \otimes (|\xi_2| \wedge n i^{X_2}_{A_2}) d(|\mu_1| \otimes |\mu_2|) \leqslant \int |\xi_1 \otimes \xi_2| d|\mu_1 \otimes \mu_2| .$$

By a), Proposition 3.7.7 a), and Corollary 3.1.2 a)

$$((|\xi_1| \wedge n i^{X_1}_{A_1}) \otimes (|\xi_2| \wedge n i^{X_2}_{A_2}))_{(A_1, A_2, n) \in \underline{\underline{R}}_1 \times \underline{\underline{R}}_2 \times \mathbb{N}}$$

is an upper directed family in M^ρ whose supremum is $|\xi_1| \tilde{\otimes} |\xi_2|$ and therefore

$$(\sup_{(A_1, n) \in \underline{\underline{R}}_1 \times \mathbb{N}} \int |\xi_1| \wedge n i^{X_1}_{A_1} d|\mu_1|)(\sup_{(A_2, n) \in R_2 \times \mathbb{N}} \int |\xi_2| \wedge n i^{X_2}_{A_2} d|\mu_2|) =$$

$$= \int |\xi_1 \otimes \xi_2| d|\mu_1 \otimes \mu_2| .$$

Since

$$\int|\xi_1 \otimes \xi_2|d|\mu_1 \otimes \mu_2| \neq 0$$

we get

$$\sup_{(A_1,n)\in \underline{\underline{R}}_1\times\mathbb{N}} \int|\xi_1|\wedge n i_{A_1}^{X_1} d|\mu_1| < \infty ,$$

$$\sup_{(A_2,n)\in \underline{\underline{R}}_2\times\mathbb{N}} \int|\xi_2|\wedge n i_{A_2}^{X_2} d|\mu_2| < \infty$$

and therefore (Theorem 3.2.4 f)) $(\xi_1,\xi_2)\in\hat{L}^1(\mu_1)\times\hat{L}^1(\mu_2)$.

d) follows immediately from Proposition 3.7.5 and Theorem 3.7.6 b_1). □

<u>*Proposition 3.7.9*</u> *Let N_1,N_2 be fundamental solid subspaces of M_1 and M_2 respectively and let F_1, F_2 be subsets of N_1^π and N_2^π respectively. Then $\xi_1 \otimes \xi_2 \in (N_1 \bar{\otimes} N_2)^\pi$ for any $(\xi_1,\xi_2)\in F_1 \times F_2$ and the topology of $(N_1,F_1) \otimes (N_2,F_2)$ is generated by the family $(q_{\xi_1\otimes\xi_2}|N_1 \otimes N_2)_{(\xi_1,\xi_2)\in F_1\times F_2}$ of seminorms on $N_1 \otimes N_2$.*

By Proposition 3.7.8 d) $\xi_1 \otimes \xi_2\in(N_1\bar{\otimes}N_2)^\pi$ for any $(\xi_1,\xi_2)\in F_1 \times F_2$.

Let G_1, G_2 be the solid subspaces of N_1^π and N_2^π generated by F_1 and F_2 respectively. By Proposition 3.4.2 a) b) G_1, G_2 are the duals of (N_1,F_1) and (N_2,F_2) respectively and the equicontinuous sets of G_1 , G_2 are exactly the order bounded sets of G_1 and G_2 respectively.

Let $(\xi_1,\xi_2)\in G_1 \times G_2$. By Theorem 3.7.6 b_1) the map

$$N_1 \otimes N_2 \longrightarrow \mathbb{R} , \ \mu \longmapsto \int\xi_1 \otimes \xi_2 d\mu$$

is the canonical linear map associated to the bilinear map

$$N_1 \times N_2 \longrightarrow \mathbb{R} , \ (\mu_1,\mu_2) \longmapsto (\int\xi_1 d\mu_1)(\int\xi_2 d\mu_2) .$$

Hence for any $(\xi_1,\xi_2)\in F_1 \times F_2$ the set

$$U(\xi_1,\xi_2) := \{\mu\in N_1\otimes N_2 \mid \forall(\eta_1,\eta_2)\in G_1 \times G_2 ,$$

$$|\eta_1| \leq |\xi_1| , |\eta_2| \leq |\xi_2| \Longrightarrow \int\eta_1 \otimes \eta_2 d\mu \leq 1\}$$

is a 0-neighbourhood in $(N_1,F_1) \otimes (N_2,F_2)$ and

$$\{U(\xi_1,\xi_2) \mid (\xi_1,\xi_2) \in F_1 \times F_2\}$$

is a fundamental system of 0-neighbourhoods in $(N_1,F_1) \otimes (N_2,F_2)$. Since for any $(\xi_1,\xi_2) \in F_1 \times F_2$ we have $|\xi_1 \otimes \xi_2| = |\xi_1| \otimes |\xi_2|$ (Theorem 3.7.6 a)) we get

$$U(\xi_1,\xi_2) = \{\mu \in N_1 \otimes N_2 \mid \int |\xi_1 \otimes \xi_2| d|\mu| \leq 1\}$$

and this shows that $(q_{\xi_1 \otimes \xi_2} | N_1 \otimes N_2)_{(\xi_1,\xi_2) \in F_1 \times F_2}$ is a family of seminorms on $N_1 \otimes N_2$ generating the topology of $(N_1,F_1) \otimes (N_2,F_2)$.□

<u>*Theorem 3.7.10*</u> *Let N, N_1, N_2 be fundamental solid subspaces of M, M_1, and M_2 respectively such that $N_1 \otimes N_2 \subset N$, let F_1, F_2 be solid subspaces of N_1^π and N_2^π respectively generating the corresponding spaces as bands and such that*

$$F := \{\xi_1 \otimes \xi_2 \mid (\xi_1 \ \xi_2) \in F_1 \times F_2\} \subset N^\pi$$

and let G be the solid subspace of N^π generated by F. Then:

a) F generates N^π as band;

b) the closure of

$$\{\sum_{\iota \in I} \mu_{1\iota} \otimes \mu_{2\iota} \mid (\mu_{1\iota},\mu_{2\iota})_{\iota \in I} \text{ finite family in } N_{1+} \times N_{2+}\}$$

in (N,F) is N_+ ;

c) $(N_1,F_1) \otimes (N_2,F_2)$ is a dense subspace of (N,F) ;

d) if (N,F) is complete (this happens if $\{1_A \mid A \in \underline{\underline{R}}\} \subset G$) there exists a unique isomorphism of locally convex spaces

$$(N_1,F_1) \tilde{\otimes} (N_2,F_2) \longrightarrow (N,F)$$

whose restriction to $N_1 \otimes N_2$ is the inclusion map $N_1 \otimes N_2 \longrightarrow N$.

a) follows from Proposition 3.7.8 b).

b) follows from a) and Proposition 3.7.2.

c) By Proposition 3.7.9 $(N_1,F_1) \otimes (N_2,F_2)$ is a subspace of (N,F) and by b) it is dense.

d) follows immediately from a), c), and Proposition 3.4.3 a). □

Corollary 3.7.11 *We set*

$$F := \{i^{X_1}_{A_1} \otimes i^{X_2}_{A_2} \mid (A_1,A_2) \in \underline{\underline{R}}_1 \times \underline{\underline{R}}_2\}$$

$$F_1 := \{i^{X_1}_{A_1} \mid A_1 \in \underline{\underline{R}}_1\} \;,\; F_2 := \{i^{X_2}_{A_2} \mid A_2 \in \underline{\underline{R}}_2\} .$$

Then:

a) $(M_1,F_1) \otimes (M_2,F_2)$ *is a dense subspace of* (M,F) *;*

b) there exists a unique isomorphism of locally convex spaces

$$(M_1,F_1) \tilde{\otimes} (M_2,F_2) \longrightarrow (M,F)$$

whose restriction to $M_1 \otimes M_2$ *is the inclusion map* $M_1 \otimes M_2 \rightarrow M$.

Let G , G_1 , G_2 be the solid subspaces of M^π , M_1^π , and M_2^π generated by F , F_1 , and F_2 respectively and let H be the set

$$\{\xi_1 \otimes \xi_2 \mid (\xi_1,\xi_2) \in G_1 \times G_2\} .$$

Then

$$(M,F) = (M,G) = (M,H) \;,\; (M_1,F_1) = (M_1,G_1) \;,\; (M_2,F_2) = (M_2,G_2) .$$

a) follows from the above relations and Theorem 3.7.10 c)

b) follows from the above relations and Theorem 3.7.10 d). □

Corollary 3.7.12 *Let* Σ , Σ_1 , Σ_2 *be the set of subsets of* $\underline{\underline{R}}$, $\underline{\underline{R}}_1$, $\underline{\underline{R}}_2$ *respectively which are closed with respect to countable unions. We set for any* $\underline{\underline{S}} \in \Sigma$, $\underline{\underline{S}}_1 \in \Sigma_1$, $\underline{\underline{S}}_2 \in \Sigma_2$

$$1_{\underline{\underline{S}}} := \bigvee_{A \in \underline{\underline{S}}} i^X_A \;,\; 1_{\underline{\underline{S}}_1} := \bigvee_{A_1 \in \underline{\underline{S}}_1} i^{X_1}_{A_1} \;,\; 1_{\underline{\underline{S}}_2} := \bigvee_{A_2 \in \underline{\underline{S}}_2} i^{X_2}_{A_2}$$

and denote

$$F := \{1_{\underline{\underline{S}}} | \underline{\underline{S}} \in \Sigma\} , \quad F_1 := \{1_{\underline{\underline{S}}_1} | \underline{\underline{S}}_1 \in \Sigma_1\}$$

$$F_2 := \{1_{\underline{\underline{S}}_2} | \underline{\underline{S}}_2 \in \Sigma_2\} .$$

Then:

a) $(M_1, F_1) \otimes (M_2, F_2)$ *is a subspace of* (M, F) ;

b) *there exists a unique isomorphism of locally convex spaces*

$$(M_1, F_1) \tilde{\otimes} (M_2, F_2) \longrightarrow (M, F)$$

whose restriction to $M_1 \otimes M_2$ *is the inclusion map* $M_1 \otimes M_2 \longrightarrow M$.

Let G , G_1 , G_2 be the solid subspaces of M^π , M_1^π , and M_2^π generated by F , F_1 , and F_2 respectively and let H be the set

$$\{\xi_1 \otimes \xi_2 | (\xi_1, \xi_2) \in G_1 \times G_2\}$$

Then

$$(M, F) = (M, G) = M(H) , \quad (M_1, F_1) = (M_1, G_1) , \quad (M_2, F_2) = (M_2, G_2).$$

a) follows from the above relations and Theorem 3.7.10 c).

b) follows from the above relations and Theorem 3.7.10 d). □

Corollary 3.7.13

a) M_{b+} *is the closure of*

$$\{ \sum_{\iota \in I} \mu_{1\iota} \otimes \mu_{2\iota} | (\mu_{1\iota}, \mu_{2\iota})_{\iota \in I} \text{ finite family in } M_{1b+} \times M_{2b+}\}$$

in $(M_b, \{i_X^X\})$;

b) $(M_{1b}, \{i_{X_1}^{X_1}\}) \otimes (M_{2b}, \{i_{X_2}^{X_2}\})$ *is a dense subspace of* $(M_b, \{i_X^X\})$;

c) *there exists a unique isomorphism of locally convex spaces*

$$(M_{1b}, \{i_{X_1}^{X_1}\}) \tilde{\otimes} (M_{2b}, \{i_{X_2}^{X_2}\}) \longrightarrow (M_b, \{i_X^X\})$$

whose restriction to $M_{1b} \otimes M_{2b}$ *is the inclusion map* $M_{1b} \otimes M_{2b} \rightarrow M_b$.

M_b^π , M_{1b}^π , M_{2b}^π are generated as solid subspaces by $\{i_X^X\}$, $\{i_{X_1}^{X_1}\}$, and $\{i_{X_2}^{X_2}\}$ respectively and by Theorem 3.7.6 c)

$$i_X^X = i_{X_1}^{X_1} \otimes i_{X_2}^{X_2} \quad .$$

By these considerations we may conclude :

a) follows from Theorem 3.7.10 b) ;

b) follows from Theorem 3.7.10 c) ;

c) follows from Theorem 3.3.1 and Theorem 3.7.10 d). □

8. The strong D.-P.-property

Definition 3.8.1 A subset A of a uniform space Y is called pseudo-compact if any sequence in A either contains a Cauchy subsequence or possesses an adherent point in Y .

Any relatively countably compact set is pseudo-compact. Each pseudo-compact set is precompact and the converse holds if the uniform space is metrizable.

Definition 3.8.2 Let E be a Hausdorff locally convex space, let E' be its strong dual and E" be its bidual, let $\underline{\underline{V}}$ be the set of convex circled 0-neighbourhoods in E the polars in E' of which are $\sigma(E',E")$-compact, and let $\underline{\underline{T}}$ be the topology on E for which $\underline{\underline{V}}$ is a fundamental system of 0-neighbourhoods in E . We say that E possesses the strong D.-P.-property if any weakly pseudo-compact set of E is precompact with respect to $\underline{\underline{T}}$. A locally convex lattice possesses the strong D.-P.-property if its underlying locally convex space possesses this property.

The strong D.-P.-property implies the D.-P.-property and the strict D.-P.-property introduced by Grothendieck ([14] Definitions 1 and 2).

Lemma 3.8.3 Let E be a locally convex space possessing the strong D.-P.-property and let φ be a continuous linear map of E into a locally convex space F . Then the following assertions are equivalent:

a) φ maps any weakly pseudo-compact set of E into a precompact set of F ;

b) φ maps any weakly compact set of E into a compact set of F ;

c) φ maps any bounded set of E into a weakly relatively compact set of F .

a ⟹ b is trivial .
b ⟹ c ⟹ a follows immediately from [14] Proposition 1, (3) ⟹ (1). □

Proposition 3.8.4 Let G be a solid subspace of M_C^{π} and let F be a subset of G. If (G,F) is Hausdorff then it possesses the strong

D.-P.-property.

By Theorem 3.4.10 d) the dual of (G,F) is a solid subspace of G^+. For any $\xi \in F$ the set

$$\{\eta \in F \mid p_\xi(\eta) \leqslant 1\}$$

is convex, solid, and (upper and lower) directed and the assertion follows immediately from [4] Theorem 4.9. □

Proposition 3.8.5 *Let* G *be a solid subspace of* M_c^π *, let* F *be a subset of* $\tilde{G}$ *such that* (G,F) *is Hausdorff, and let* φ *be a continuous linear map of* G *into a locally convex space* E *. Then the following assertions are equivalent:*

a) φ *maps any weakly pseudo-compact set of* (G,F) *into a precompact set of* E *;*

b) φ *maps any weakly compact set of* (G,F) *into a compact set of* E *;*

c) φ *maps any bounded set of* (G,F) *into a weakly relatively compact set of* E *.*

The assertion follows immediately from Proposition 3.8.4 and Lemma 3.8.3. □

Corollary 3.8.6 *Let* E *be a locally convex space, let* φ *be a linear map of* M_c^π *into* E*, continuous with respect to the Mackey topology* $\tau(M_c^\pi, M_c)$*, and let* F *be the set* $\{1_A \mid A \in \underline{\underline{R}}\}$ *. Then* φ *maps any weakly pseudo-compact set of* (M_c^π, F) *into a precompact set of* E *and maps any weakly compact set of* (M_c^π, F) *into a compact set of* E *.*

By Theorem 3.4.9 any bounded set of (M_c^π, F) is relatively compact with respect to the $\sigma(M_c^\pi, M_c)$-topology. Hence φ maps any bounded set of (M_c^π, F) into a weakly relatively compact set of E . The assertion follows from Proposition 3.8.5 c $\Longrightarrow$ a & b. □

Definition 3.8.7 *Let* E *be a vector lattice. The finest locally convex topology on* E *for which each order interval is bounded is called the order topology of* E *.*

Proposition 3.8.8 *Let* $\underline{\underline{C}}$ *be a subset of* $\underline{\underline{R}}$ *such that the sets of* $\underline{\underline{C}}$ *are pairwise disjoint and such that for any* $A \in \underline{\underline{R}}$ *there exists a finite subset* $\underline{\underline{C}}_o$ *of* $\underline{\underline{C}}$ *with* $A \subset \bigcup_{C \in \underline{\underline{C}}_o} C$*. Then:*

a) $\underline{\underline{C}}$ *is a locally finite M-concassage ;*

b) for any representation (Y,u,v) *of* (X,M) *the space* Y *is paracompact ;*

c) for any $\xi \in M^\pi$ *there exists* $A \in \underline{\underline{R}}$ *such that* $\xi \dot{1}_A = \xi$ *;*

d) M^π *endowed with the order topology possesses the strong D.-P.-property and for any bounded set* A *of* M^π *with respect to the order topology there exists* $\xi \in M^\pi$ *such that*

$$A \subset \{\eta \in M^\pi \mid |\eta| \leq \xi\} .$$

a) is trivial.

b) $(\mathrm{Supp}\, u\dot{1}_C)_{C \in \underline{\underline{C}}}$ is a disjoint family of open compact sets of Y whose union is Y. Hence Y is paracompact.

c & d. By Theorem 2.3.8 there exists a representation of (X,M) and by Proposition 3.1.1 a), b), c) the map

$$M^\pi \longrightarrow C_i(Y) \ , \ \xi \longmapsto \hat{u}\xi$$

is an isomorphism of vector lattices. By b) Y is paracompact and therefore $C_i(Y)$ is the set of continuous real functions on Y with compact carrier. c) follows immediately from these considerations as well as the second assertion of d). The first assertion of d) follows from the above considerations and from [4] Corollary 4.14. □

9. The strong approximation property

Definition 3.9.1 *Let* E *be a normed space, let* L(E) *be the normed vector space of continuous linear maps of* E *into itself, and let* A *be the set of* $u \in L(E)$ *such that* $\|u\| \leq 1$ *and such that* u(E) *is of finite dimension.* E *possesses the metric approximation property if the identity map of* E *is an adherent point of* A *with respect to the topology of precompact convergence* (A. Grothendieck [15] Definition I 10).

We may replace the topology of precompact convergence in the above definition by the topology of pointwise convergence in a dense set of E.

Proposition 3.9.2 *Any solid subspace of the Banach lattice* M_b *possesses the metric approximation property.*

Let N be a solid subpsace of M_b. We denote by Δ the set of finite subsets $\underline{\underline{A}}$ of $\underline{\underline{R}}$ such that the sets of $\underline{\underline{A}}$ are pairwise disjoint and set

$$P := \{\mu \in N_+ \mid \|\mu\| = 1\} ,$$

$$U := \{ \sum_{A \in \underline{\underline{A}}} i_A \otimes \mu_A \mid \underline{\underline{A}} \in \Delta , \quad (\mu_A)_{A \in \underline{\underline{A}}} \text{ family in } P\} .$$

Let $\underline{\underline{A}} \in \Delta$ and let $(\mu_A)_{A \in \underline{\underline{A}}}$ be a family in P. We have

$$\|(\sum_{A \in \underline{\underline{A}}} i_A \otimes \mu_A)(\mu)\| = \| \sum_{A \in \underline{\underline{A}}} \mu(A) \mu_A\| \leq$$

$$\leq \sum_{A \in \underline{\underline{A}}} |\mu(A)| \, \|\mu_A\| \leq \|\mu\|$$

for any $\mu \in N$ and therefore

$$\| \sum_{A \in \underline{\underline{A}}} i_A \otimes \mu_A\| \leq 1.$$

We want to show that the identity map of N is an adherent point of

$\mathcal{U}$ with respect to the topology of pointwise convergence in a dense set of $\mathcal{N}$.

Let $(\mu_\iota)_{\iota\in I}$ be a family in $\mathcal{N}_+$ generating $\mathcal{N}$ as band and such that

$$\forall \iota, \iota' \in I, \iota \neq \iota' \Longrightarrow \mu_\iota \wedge \mu_{\iota'} = 0 .$$

Let further Λ be the set of pairs $(J, \underline{\underline{A}})$ such that :

a) J is a finite subset of I ;

b) $\underline{\underline{A}} \in \Delta$;

c) $\forall A \in \underline{\underline{A}}, \forall \iota, \iota' \in J, \iota \neq \iota' \Longrightarrow \mu_\iota(A)\mu_{\iota'}(A) = 0.$

For any $(J, \underline{\underline{A}}), (J', \underline{\underline{A}}') \in \Lambda$ we set $(J, \underline{\underline{A}}) \leqslant (J', \underline{\underline{A}}')$ if :

a) $J \subset J'$;

b) $\bigcup_{A \in \underline{\underline{A}}} A \subset \bigcup_{A' \in \underline{\underline{A}}'} A'$;

c) $\forall (A, A') \in \underline{\underline{A}} \times \underline{\underline{A}}', A \cap A' \neq \emptyset \Longrightarrow A' \subset A$.

$\leqslant$ is an upper directed order relation on Λ. We denote by $\underline{\underline{G}}$ the section filter of Λ. We set for any $\iota \in I$:

$$\underline{\underline{R}}_\iota := \{A \in \underline{\underline{R}} \mid \mu_\iota(A) > 0\} ,$$

$$\mathcal{N}_\iota := \{\nu \in \mathcal{M}_c \cap \mathcal{N} \mid \exists n \in \mathbb{N}, |\nu| \leqslant n\mu_\iota\} .$$

We denote by $\mathcal{L}(\mathcal{N})$ the vector space of continuous linear maps of $\mathcal{N}$ into itself and by ψ the map

$$\Lambda \longrightarrow \mathcal{L}(\mathcal{N}), \quad (J, \underline{\underline{A}}) \longmapsto \sum_{\iota \in J} \sum_{A \in \underline{\underline{A}} \cap \underline{\underline{R}}_\iota} \frac{1}{\mu_\iota(A)} (\dot{1}_A \otimes (1_A \cdot \mu_\iota))$$

and set $\underline{\underline{F}} := \psi(\underline{\underline{G}})$. Since $\psi(\Lambda) \subset \mathcal{U}$ we have $\mathcal{U} \in \underline{\underline{F}}$.

Let $\iota \in I$ and let $\nu \in \mathcal{N}_\iota$. We want to prove

$$\lim_{u, \underline{\underline{F}}} u(\nu) = \nu .$$

Since $\nu \in M_c$ there exists $B \in \underline{\underline{R}}$ such that $X \setminus B$ is a ν-null set. By Radon-Nikodym theorem there exists an $\underline{\underline{R}}$-measurable bounded real function f on X such that $\nu = f \cdot \mu_\iota$. Let ε be a strictly positive real number. There exists $\underline{\underline{A}}_0 \in \Delta$ such that $\bigcup_{A \in \underline{\underline{A}}_0} A = B$ and such that the oscillation of f on each set of $\underline{\underline{A}}_0$ is smaller that $\frac{\varepsilon}{1+\|\mu_\iota\|}$

Then $(\{\iota\}, \underline{\underline{A}}_0) \in \Lambda$. Let $(J,\underline{\underline{A}}) \in \Delta$ with

$$(\{\iota\}, \underline{\underline{A}}_0) \leqslant (J,\underline{\underline{A}}) \ .$$

We have $\mu_\iota(A) = 0$ and therefore $|\nu|(A) = 0$, $1_A \cdot \nu = 0$, and $1_A \cdot \mu_\iota = 0$ for any $A \in \underline{\underline{A}} \setminus \underline{\underline{R}}_\iota$. Since $B \subset \bigcup_{A \in \underline{\underline{A}}} A$ we get

$$\nu = 1_B \cdot \nu = \sum_{A \in \underline{\underline{A}} \cap \underline{\underline{R}}_\iota} 1_A \cdot \nu \ ,$$

$$(\psi(J,\underline{\underline{A}}))(\nu) = \sum_{A \in \underline{\underline{A}} \cap \underline{\underline{R}}_\iota} \frac{\nu(A)}{\mu_\iota(A)} 1_A \cdot \mu_\iota \ .$$

We have

$$\left\|1_A \cdot \nu - \frac{\nu(A)}{\mu_\iota(A)} 1_A \cdot \mu_\iota\right\| = \int_A \left| f - \frac{1}{\mu_\iota(A)} \int_A f d\mu_\iota \right| d\mu_\iota \leqslant$$

$$< \frac{\varepsilon}{1+\|\mu_\iota\|} \mu_\iota(A)$$

for any $A \in \underline{\underline{A}} \cap \underline{\underline{R}}_\iota$ and therefore

$$\|\nu - (\psi(J,A))(\nu)\| = \left\| \sum_{A \in \underline{\underline{A}} \cap \underline{\underline{R}}_\iota} 1_A \cdot \nu - \sum_{A \in \underline{\underline{A}} \cap \underline{\underline{R}}_\iota} \frac{\nu(A)}{\mu_\iota(A)} 1_A \cdot \mu_\iota \right\| \leqslant$$

$$\leqslant \sum_{A \in \underline{\underline{A}} \cap \underline{\underline{R}}_\iota} \left\|1_A \cdot \nu - \frac{\nu(A)}{\mu_\iota(A)} 1_A \cdot \mu_\iota\right\| \leqslant \frac{\varepsilon}{1+\|\mu_\iota\|} \|\mu_\iota\| < \varepsilon \ .$$

ε being arbitrary we get

$$\lim_{u,\underline{\underline{F}}} u(\nu) = \nu \ .$$

Since ι is arbitrary it follows from the above considerations that

$$\lim_{u,\underline{\underline{F}}} u(\nu) = \nu$$

for any ν belonging to the vector subspace of N generated by $\bigcup_{\iota \in I} N_\iota$, hence for any ν belonging to a dense set of N. □

<u>*Definition 3.9.3*</u> *Let* E *be a locally convex space and let* U *be a circled convex 0-neighbourhood in* E *. We set* $F := \bigcap_{n \in \mathbb{N}} \frac{1}{n} U$ *and denote by* E_U *the vector space* E/F *endowed with the norm*

$$E/F \longrightarrow \mathbb{R}_+ , \ x \longmapsto \inf \{\alpha \in \mathbb{R}_+ | x \in \alpha U\} .$$

We say that E *possesses the strong approximation property if there exists a fundamental system* $\underline{\underline{U}}$ *of closed circled convex 0-neighbourhoods in* E *such that* E_U *possesses the metrique approximation property for any* $U \in \underline{\underline{U}}$.

<u>*Definition 3.9.4*</u> *Let* E *be a locally convex space, let* $L(E)$ *be the vector space of continuous linear maps of* E *into itself, and let* $L_f(E)$ *be the set of* $u \in L(E)$ *for which* $u(E)$ *is finite dimensional.* E *possesses the approximation property if the identity map of* E *is an adherent point of* $L_f(E)$ *with respect to the topology of precompact convergence.* (A. Grothendieck [15] Definition I 9).

The following result will not be used in the sequel, it shows only the relation between the strong approximation property and the approximation property.

<u>*Lemma 3.9.5*</u> *Let* E *be a locally convex space possessing the strong approximation property, let* $L(E)$ *be the set of continuous linear maps of* E *into itself, and let* $L_f(E)$ *be the set of* $u \in L(E)$ *for which* $u(E)$ *is finite dimensional. Then there exists a filter* $\underline{\underline{F}}$ *on* $L(E)$ *converging to the identity map of* E *with respect to the topology of precompact convergence, containing* $L_f(E)$*, and such that for any 0-neighbourhood* U *in* E *there exists* $\mathcal{U} \in \underline{\underline{F}}$ *such that* $\bigcap_{u \in \mathcal{U}} u^{-1}(U)$ *is a 0-neighbourhood in* E *. In particular* E *possesses the approxima-*

tion property.

Let $\underline{\underline{U}}$ be a fundamental system of closed circled convex 0-neighbourhoods in E such that E_U possesses the metric approximation property for any $U \in \underline{\underline{U}}$.

Let $U \in \underline{\underline{U}}$, let u be the canonical map $E \longrightarrow E_U$, let A be precompact set of E , and let ε be a strictly positive real number. Then $u(A)$ is a precompact set of E_U and therefore there exists a finite family $((y_\iota , y'_\iota))_{\iota \in I}$ in $E_U \times E'_U$ such that

$$\| y - \sum_{\iota \in I} < y, y'_\iota > y_\iota \| \leq \varepsilon$$

for any $y \in u(A)$ and

$$\| \sum_{\iota \in I} y_\iota \otimes y'_\iota \| \leq 1 .$$

Let $(x_\iota)_{\iota \in I}$ be a family in E such that $y_\iota = u(x_\iota)$ for any $\iota \in I$. We set

$$v := \sum_{\iota \in I} x_\iota \otimes (y'_\iota \circ \mu).$$

Then $v \in L_f(E)$ and

$$u\,(v(x)) = u(\sum_{\iota \in I} y'_\iota (u(x)) x_\iota) = \sum_{\iota \in I} y'_\iota (u(x)) y_\iota$$

for any $x \in E$. We get $u(v(x)) \in u(U)$ for any $x \in U$ and

$$u(v(x) - x) \in u(\varepsilon U)$$

for any $x \in A$. Hence $U \subset \overset{-1}{v}(U)$ and $v(x) - x \in \varepsilon U$ for any $x \in A$.

We denote by Λ the set of elements $(U, \underline{\underline{V}}, A, \varepsilon)$ such that $U \in \underline{\underline{U}}$, $\underline{\underline{V}}$ is a finite subset of $\underline{\underline{U}}$ with $U \subset \bigcap_{V \in \underline{\underline{V}}} V$, A is a precompact set of E, and ε is a strictly positive real number. We denote for any $(U, \underline{\underline{V}}, A, \varepsilon) \in \Lambda$ by $\mathcal{U}(U, \underline{\underline{V}}, A, \varepsilon)$ the set of $u \in L_f(E)$ such that

$$U \subset \bar{u}^{1}(\bigcap_{V \in \underline{\underline{V}}} V) \quad \text{and}$$

$$u(x) - x \in \varepsilon U$$

for any $x \in A$. By the above considerations $\mathcal{U}(U,\underline{\underline{V}},A,\varepsilon)$ is nonempty for any $(U,\underline{\underline{V}},A,\varepsilon) \in \Lambda$. It is easy to see that

$$\{\mathcal{U}(U,\underline{\underline{V}},A,\varepsilon) \mid (U,\underline{\underline{V}},A,\varepsilon) \in \Lambda\}$$

is a filter base. The filter on $L(E)$ generated by this filter base possesses the required properties. □

<u>*Theorem 3.9.6*</u> *Let N be a fundamental solid subspace of M and let F be a subset of N^{π} . Then (N,F) possesses the strong approximation property.*

Let $\xi \in F$. We set

$$N_{\xi} := \{\xi \cdot \mu \mid \mu \in N\}$$

Then N_{ξ} is a solid subspace of M_b (Theorems 3.2.2 and 3.1.7). Let us endow N_{ξ} with the norm induced by M_b . By Proposition 3.9.2 N_{ξ} possesses the metric approximation property. Let N_{ξ}^{o} be the canonical vector space associated to N and q_{ξ} (i.e. $N_{\xi}^{o} := N/\overset{-1}{q_{\xi}}(o)$) . By Theorem 3.2.2 N_{ξ}^{o} is isomorphic to N_{ξ} as normed space and therefore it possesses the metric approximation property.

Since ξ is arbitrary N possesses the strong approximation property. □

§ 4 VECTOR MEASURES

Throughout this chapter we denote by E *a (Hausdorff) locally convex space* .

We denote for any locally convex space F by F' its strong dual and by F" its bidual and for any real vector space F by F* its algebraic dual. If <F,G> is a duality we denote by F_G the space F endowed with the weak topology $\sigma(F,G)$ associated to this duality.

1. Preliminaries

<u>*Definition 4.1.1*</u> *A δ-filter is a filter* $\underline{\underline{F}}$ *for which the intersection of any sequence in* $\underline{\underline{F}}$ *belongs to* $\underline{\underline{F}}$. *A filter* $\underline{\underline{F}}$ *on a locally convex space* F *is called* bounded *if it contains a bounded set of* F . *A locally convex space is called δ-quasicomplete if any bounded Cauchy δ-filter on it converges.*

Any quasicomplete locally convex space and any weakly δ-quasicomplete space is δ-quasicomplete. Any locally convex space is δ-quasicomplete if its one-point sets are G_δ-sets.

<u>*Definition 4.1.2*</u> *A sequence* $(x_n)_{n\in\mathbb{N}}$ *in a locally convex space* F *will be called a* Σ-sequence *if for any* 0*-neighbourhood* U *in* F *there exists a finite subset* I *of* $\mathbb{N}$ *such that*

$$\sum_{n\in J}(x_{n+1}-x_n)\in U$$

for any finite subset J *of* $\mathbb{N}\setminus I$. *A locally convex space will be called* Σ-complete *if any Σ-sequence in this space converges.*

Any Σ-sequence is a Cauchy sequence and so any sequentially complete locally convex space and any weakly Σ-complete locally convex space is Σ-complete. Any Cauchy sequence in a metrizable locally convex space possesses a Σ-subsequence and therefore for such spaces Σ-completeness and completeness coincide. Any semireflexive locally convex space is weakly Σ-complete.

<u>*Lemma 4.1.3*</u> *Let* (I,f) *be a net in* E *(i.e.* I *is an upper direct-*

ed preordered set and f a map of I into E) such that $(f(\iota_n))_{n\in\mathbb{N}}$ *is a Cauchy sequence for any increasing sequence* $(\iota_n)_{n\in\mathbb{N}}$ *in* I *and let* $\underline{\underline{F}}$ *be the section filter of* I *. Then* $f(\underline{\underline{F}})$ *is a Cauchy filter on* E.

Let U be a convex 0-neighbourhood in E. Assume for any $A\in f(\underline{\underline{F}})$ there exists $x,y\in A$ with $x - y\notin U$. Then we may construct inductively an increasing sequence $(\iota_n)_{n\in\mathbb{N}}$ in I such that

$$m \neq n \Longrightarrow f(\iota_m) - f(\iota_n)\notin U$$

an this is a contradiction. Hence $f(\underline{\underline{F}})$ is a Cauchy filter on E .□

Lemma 4.1.4 *Let* $(x_n)_{n\in\mathbb{N}}$ *be a sequence in* E *. If the sequence* $(\sum_{\substack{m\in M\\ m\leq n}} x_m)_{n\in\mathbb{N}}$ *converges for any* $M\subset\mathbb{N}$ *then the family* $(x_n)_{n\in M}$ *is summable for any* $M\subset\mathbb{N}$.

It is sufficient to show that $(x_n)_{n\in\mathbb{N}}$ is summable. Let $\underline{\underline{I}}$ be the set of finite subsets of $\mathbb{N}$ ordered by the inclusion relation, let $\underline{\underline{F}}$ be its section filter, and let φ be the map

$$\underline{\underline{I}} \longrightarrow E \, , \, I \longmapsto \sum_{n\in I} x_n \, .$$

Let V be a 0-neighbourhood in E. Assume for any $I\in\underline{\underline{I}}$ there exists $J\in\underline{\underline{I}}$ with $I\cap J = \emptyset$ and $\varphi(J)\notin V$. We may construct inductively a sequence $(I_n)_{n\in\mathbb{N}}$ in $\underline{\underline{I}}$ such that

$$p\in I_n \, , \, q\in I_{n+1} \Longrightarrow p < q$$

$$\varphi(I_n)\notin V$$

for any $n\in\mathbb{N}$. We set $M := \bigcup_{n\in\mathbb{N}} I_n$. The above relations contradict the hypothesis that $(\sum_{\substack{m\in M\\ m\leq n}} x_m)_{n\in\mathbb{N}}$ converges. Hence $\varphi(\underline{\underline{F}})$ is a Cauchy filter. Since $\lim_{n\to\infty} \sum_{\substack{m\in M\\ m\leq n}} x_m$ is an adherent point of $\varphi(\underline{\underline{F}})$, $\varphi(\underline{\underline{F}})$ converges.□

Definition 4.1.5 A linear map u of a vector lattice F into E will be called order continuous if for any lower directed nonempty set A in F with infimum 0, $u(\underline{\underline{F}})$ converges to 0 where $\underline{\underline{F}}$ denotes the filter on F generated by the filter base

$$\{\{y \in A \mid y \leqslant x\} \mid x \in A\} \quad .$$

It is easy to see that this definition reduces to the older one (page 5) if $E = R$.

Proposition 4.1.6 ([5] Theorem 9). *Let F be an order σ-complete vector lattice and let φ be a linear map of F into E . If φ is order continuous with respect to the weak topology of E then it is order continuous with respect to the initial topology of E .*

Let A be a lower directed nonempty subset of F with infimum 0 and let $\underline{\underline{F}}$ be the filter on F generated by

$$\{\{y \in A \mid y \leqslant x\} \mid x \in A\} \quad .$$

We have to show that $\varphi(\underline{\underline{F}})$ converges to 0 . Since it weakly converges to 0 it is sufficient to show that $\varphi(\underline{\underline{F}})$ is a Cauchy filter.

Let $(x_n)_{n \in \mathbb{N}}$ be a decreasing sequence in A. Let M be a subset of $\mathbb{N}$. Then

$$\Big(\sum_{\substack{n \in M \\ n \leqslant m}} (x_n - x_{n+1})\Big)_{m \in \mathbb{N}}$$

is an increasing upper bounded sequence in F ; since F is order σ-complete it possesses a supremum in F. Using the hypothesis that φ is order σ-continuous with respect to the weak topology of E we deduce that the sequence

$$\Big(\sum_{\substack{n \in M \\ n \leqslant m}} (x_n - x_{n+1})\Big)_{m \in \mathbb{N}}$$

is weakly convergent. By Lemma 4.1.4 $((x_n - x_{n+1}))_{n \in M}$ is weakly summable for any $M \subset N$. By Orlicz-Pettis theorem $((x_n - x_{n+1}))_{n \in \mathbb{N}}$

is summable with respect to the initial topology of E . Hence $((x_n))_{n\in\mathbb{N}}$ is a Cauchy sequence in E . By Lemma 4.1.3 $\varphi(\underline{\underline{F}})$ is a Chauchy filter. □

2. The integral with respect to a vector measure

Definition 4.2.1 *We denote for any locally convex space* F *and for any set* N *of measures on* $\underline{\underline{R}}$ *by* $N(F)$ *the set of maps of* $\underline{\underline{R}}$ *into* F *such that* $x'\circ\mu\in N$ *for any* $x'\in F'$.

Let $\mu\in M(E)$ and let $(A_\iota)_{\iota\in I}$ be a countable family of pairwise disjoint sets of $\underline{\underline{R}}$ the union of which belongs to $\underline{\underline{R}}$. Then $(\mu(A_\iota))_{\iota\in I}$ is a summable family in E with respect to the weak topology of E and its sum is $\mu(\bigcup_{\iota\in I} A_\iota)$. By Orlicz-Pettis theorem $(\mu(A_\iota))_{\iota\in I}$ is a summable family in E with respect to the initial topology of E and its sum is $\mu(\bigcup_{\iota\in I} A_\iota)$. Hence μ is an E-valued measure on $\underline{\underline{R}}$. In particular if $M = M(\underline{\underline{R}})$ then $M(E)$ is the set of all E-valued measures on $\underline{\underline{R}}$.

Let $\mu\in M_b(E)$. Then $\mu(\underline{\underline{R}})$ is a weakly bounded and therefore a bounded set of E . Conversely if $\mu(\underline{\underline{R}})$ is a bounded set of E for a $\mu\in M(E)$ then $\mu\in M_b(E)$. Hence $M_b(E)$ is the set of $\mu\in M(E)$ for which $\mu(\underline{\underline{R}})$ is bounded.

For any subspace N of M , $N(E)$ is a subspace of $M(E)$.

The following result will not be used in the sequel it has only the mission to clarify the notion $M(E)$.

Proposition 4.2.2 *Let* $\underline{\underline{K}}$ *be a subset of* $\underline{\underline{R}}$ *closed with respect to finite unions. We assume a real measure* λ *on* $\underline{\underline{R}}$ *belongs to* M *iff for any* $A\in\underline{\underline{R}}$ *and for any* $\varepsilon > 0$ *there exists* $K\in\underline{\underline{K}}$ *such that* $K\subset A$ *and*

$$|\lambda(B) - \lambda(A)| < \varepsilon$$

for any $B\in\underline{\underline{R}}$ *with* $K\subset B\subset A$. *Then an* E-*valued measure* μ *on* $\underline{\underline{R}}$ *belongs to* $M(E)$ *iff for any* $A\in\underline{\underline{R}}$ *and for any* 0-*neighbourhood* U *in* E *there exists* $K\in\underline{\underline{K}}$ *such that* $K\subset A$ *and*

$$\mu(B) - \mu(A) \in U$$

for any $B \in \underline{\underline{R}}$ *with* $K \subset B \subset A$.

The sufficiency is trivial. Let us prove the necessity. Let $\mu \in M(E)$, let $A \in \underline{\underline{R}}$, let I be the set $\{K \in \underline{\underline{K}} \mid K \subset A\}$ ordered by the inclusion relation, and let $\underline{\underline{F}}$ be the section filter of I . We want to show that $\mu(\underline{\underline{F}})$ is a Cauchy filter on E . Assume the contrary. Then there exists a 0-neighbourhood U in E such that for any $K \in I$ there exists $K' \in I$ with $K \subset K'$ and

$$\mu(K') - \mu(K) \notin U .$$

We may construct inductively an increasing sequence $(K_n)_{n \in \mathbb{N}}$ in I such that

$$\mu(K_{n+1} \setminus K_n) = \mu(K_{n+1}) - \mu(K_n) \notin U$$

for any $n \in \mathbb{N}$. The sequence $(\mu(K_{n+1} \setminus K_n))_{n \in \mathbb{N}}$ being summable it converges to 0 and this is the expected contradiction. Hence $\mu(\underline{\underline{F}})$ is a Cauchy filter. By the hypothesis it converges to $\mu(A)$ with respect to the weak topology of E . Hence $\mu(\underline{\underline{F}})$ converges to $\mu(A)$ with respect to the initial topology of E .

Let now V be a closed 0-neighbourhood in E . Assume for any $K \in I$ there exists $B \in \underline{\underline{R}}$ such that $K \subset B \subset A$ and

$$\mu(B) - \mu(A) \notin V .$$

We construct inductively a disjoint sequence $(K_n)_{n \in \mathbb{N}}$ in I such that $\mu(K_n) \notin V$ for any $n \in \mathbb{N}$. Let $n \in \mathbb{N}$ and assume the sequence was constructed up to $n - 1$. By the hypothesis of the proof there exists $B \in \underline{\underline{R}}$ such that

$$\bigcup_{m<n} K_m \subset B \subset A \ , \ \mu(B) - \mu(A) \notin V .$$

Then $\mu(A \setminus B) \notin V$ and by the first part of the proof there exists $K_n \in \underline{\underline{K}}$ such that $K_n \subset A \setminus B$ and $\mu(K_n) \notin V$. The existence of the sequence $(K_n)_{n \in \mathbb{N}}$ is a contradiction since $(\mu(K_n))_{n \in \mathbb{N}}$ is summable and therefore it converges to 0 . Hence there exists $K \in \mathfrak{I}$ such that

$$\mu(B) - \mu(A) \in V$$

for any $B \in \underline{\underline{R}}$ with $K \subset B \subset A$. □

Remark. Assume X is a Hausdorff topological space, $\underline{\underline{K}}$ is the set of compact sets of X, and $\underline{\underline{R}}$ is the set of relatively compact Borel sets of X . Then M is the set of Radon real measures on X and the above result shows that $M(E)$ is the set of Radon E-valued measures on X. This result was proved by D.R. Lewis ([21] Theorem 1.6).

Definition 4.2.3 *Let* $\mu \in M(E)$. *We set*

$$\hat{L}^1(\mu) := \bigcap_{x' \in E'} \hat{L}^1(x' \circ \mu) \quad , \quad \hat{L}^1_{loc}(\mu) := \bigcap_{x' \in E'} \hat{L}^1_{loc}(x' \circ \mu)$$

and call the elements of $\hat{L}^1(\mu)$ *and* $\hat{L}^1_{loc}(\mu)$ μ*-integrable and locally* μ*-intergrable respectively.* $\hat{L}^1(\mu)$ *and* $\hat{L}^1_{loc}(\mu)$ *are solid subspaces of* M^ρ *(Corollary 3.1.3). For any* $\xi \in \hat{L}^1(\mu)$ *we denote by* $\int \xi d\mu$ *the map*

$$E' \longrightarrow \mathbb{R} \, , \; x' \longmapsto \int \xi d(x' \circ \mu) \quad ;$$

it belongs to the algebraic dual E'^* *of* E' *and the map*

$$\hat{L}^1(\mu) \longrightarrow E'^* \, , \; \xi \longmapsto \int \xi d(x' \circ \mu)$$

is linear. For any $\xi \in \hat{L}^1_{loc}(\mu)$ *we denote by* $\xi \cdot \mu$ *the map*

$$\underline{\underline{R}} \longrightarrow E'^* \, , \quad A \longmapsto \int \xi \dot{1}_A d\mu \; .$$

For any $f \in L_\infty$ *with* $\dot{f} \in \hat{L}^1(\mu)$ $(\dot{f} \in \hat{L}^1_{loc}(\mu))$ *we set*

$$\int f d\mu := \int \dot{f} d\mu \qquad (f \cdot \mu := \dot{f} \cdot \mu) \, .$$

Let $\mu \in M(E)$. We have

$$M^\pi_c \subset \hat{L}^1_{loc}(\mu) \quad \text{(Proposition 3.2.1)} \; ,$$

$$\dot{1} \cdot \mu = \mu \; ,$$

$$\mu \in M_b(E) \iff \dot{1} \in \hat{\mathcal{L}}(\mu) \quad ,$$

$$\forall A \in \underline{\underline{R}} \implies \dot{1}_A \cdot \mu \in M_b(E) \quad ,$$

$$\forall (\xi, x') \in \hat{\mathcal{L}}^1_{loc}(\mu) \times E' \implies x' \circ (\xi \cdot \mu) = \xi \cdot (x' \circ \mu) \quad .$$

Proposition 4.2.4 *Let* $\mu \in M(E)$, *let* $\eta \in \hat{\mathcal{L}}^1_{loc}(\mu)$ *such that* $\eta \cdot \mu \in M(E)$, *and let* $\xi \in M^\rho$. *Then:*

$$\xi \in \hat{\mathcal{L}}^1(\eta \cdot \mu) \iff \xi\eta \in \hat{\mathcal{L}}^1(\mu) \implies \int \xi d(\eta \cdot \mu) = \int \xi\eta d\mu \ ;$$

$$\xi \in \hat{\mathcal{L}}^1_{loc}(\eta \cdot \mu) \iff \xi\eta \in \hat{\mathcal{L}}^1_{loc}(\mu) \implies \xi \cdot (\eta \cdot \mu) = (\xi\eta) \cdot \mu \ ;$$

$$\eta \in \hat{\mathcal{L}}^1(\mu) \iff \eta \cdot \mu \in M_b(E) \ .$$

The assertions follow from Theorem 3.2.2 d) . □

Proposition 4.2.5 *Let* $\mu \in M_b(E)$ *and let* $\Gamma\mu(\underline{\underline{R}})$ *be the closed convex circled hull of* $\mu(\underline{\underline{R}})$ *in* $E''_{E'}$. *Then*

a) $M_b^\pi \subset \hat{\mathcal{L}}^1(\mu)$;

b) *we have*

$$\Gamma\mu(\underline{\underline{R}}) \subset \{\int \xi d\mu \mid \xi \in M_b^\pi, \ \|\xi\| \leq 1\} \subset 2\Gamma\mu(\underline{\underline{R}}) \ ;$$

c) $\{\int \xi d\mu \mid \xi \in M_b^\pi \ , \ \|\xi\| \leq 1\}$ *is a compact convex circled set of* $E''_{E'}$;

d) *if* F *is a subspace of* E *then the map*

$$\{\xi \in M_b^\pi \mid \int \xi d\mu \in F\} \longrightarrow F \ , \ \xi \longmapsto \int \xi d\mu$$

is continuous with respect to the norm topology on $\{\xi \in M_b^\pi \mid \int \xi d\mu \in F\}$ *induced by* M_b^π *and the finest locally convex topology on* F *for which* $F \cap \Gamma\mu(\underline{\underline{R}})$ *is a bounded set.*

a) is trivial.

b) Let $\mu(\underline{\underline{R}})^\circ$ be the absolute polar set of $\mu(\underline{\underline{R}})$ in E'. We have

$$|<\int \xi d\mu \ , \ x'>| = |\int \xi d(x' \circ \mu)| \leq \|\xi\| \ \|x' \circ \mu\| \leq \|\xi\|$$

for any $\xi \in M_b^\pi$ and for any $x' \in \frac{1}{2}\mu(\underline{\underline{R}})^\circ$. Hence

$$\{\int \xi d\mu \mid \xi \in M_b^{\pi} , \ \|\xi\| \leq 1\}$$

is contained in the polar set of $\frac{1}{2}\mu(\underline{\underline{R}})^{o}$ in E'^{*}. Since $\frac{1}{2}\mu(\underline{\underline{R}})$ is bounded the polar set in E'^{*} lies in E'' and is equal to $2\Gamma\mu(\underline{\underline{R}})$. We get

$$\{\int \xi d\mu \mid \xi \in M_b^{\pi} , \ \|\xi\| \leq 1\} \subset 2\Gamma\mu(\underline{\underline{R}}) .$$

The other inclusion is trivial.

c) The map

$$M_b^{\pi} \longrightarrow E'' , \ \xi \longmapsto \int \xi d\mu$$

is the adjoint of the map

$$E' \longrightarrow M_b , \ x' \longmapsto x' \circ \mu$$

and therefore it is continuous with respect to the corresponding weak topologies. Since $\{\xi \in M_b^{\pi} \mid \|\xi\| \leq 1\}$ is compact for the weak $\sigma(M_b^{\pi}, M_b)$ topology its image

$$\{\int \xi d\mu \mid \xi \in M_b^{\pi} , \ \|\xi\| \leq 1\}$$

is a compact set of $E''_{E'}$. It is obvious that this set is convex and circled.

d) follows immediately from b) . □

Definition 4.2.6 We say that a δ-ring $\underline{\underline{S}}$ is a quasi-σ-ring if any disjoint sequence in $\underline{\underline{S}}$ possesses a subsequence whose union belongs to $\underline{\underline{S}}$.

The set of finite subsets of $\mathbb{N}$ is a δ-ring which is not a quasi-σ-ring. Any σ-ring is a quasi-σ-ring. If $\underline{\underline{F}}$ is a free ultrafilter on $\mathbb{N}$ then

$$\{A \subset N \mid A \notin \underline{\underline{F}}\}$$

is a quasi-σ-ring which is not a σ-ring. The set

$$\{A \subset \mathbb{N} \mid \forall \alpha \in \mathbb{R}_+ \setminus \{0\} \Longrightarrow \sum_{n\in A} \frac{1}{n^\alpha} < \infty\}$$

is another example of a quasi-σ-ring which is not a σ-ring.

Proposition 4.2.7 *If* $\underline{\underline{R}}$ *is a quasi-σ-ring and if* $\mu \in M(E)$ *then:*

a) for any disjoint family $(A_\iota)_{\iota\in I}$ *in* $\underline{\underline{R}}$ *and for any 0-neighbourhood* U *in* E *the set* $\{\iota\in I \mid \mu(A_\iota) \notin U\}$ *is finite ;*

b) for any disjoint sequence $(A_n)_{n\in\mathbb{N}}$ *in* $\underline{\underline{R}}$ *the sequence* $(\mu(A_n))_{n\in\mathbb{N}}$ *converges to* 0 *;*

c) for any increasing sequence $(A_n)_{n\in\mathbb{N}}$ *in* $\underline{\underline{R}}$ *,* $(\mu(A_n))_{n\in\mathbb{N}}$ *is a* Σ*-sequence in* E *;*

d) $\mu \in M_b(E)$.

a) Assume the set $\{\iota\in I \mid \mu(A_\iota) \notin U\}$ is not finite. Then there exists an infinite subset J of $\{\iota\in I \mid \mu(A_\iota) \notin U\}$ with $\bigcup_{\iota\in J} A_\iota \in \underline{\underline{R}}$. Since $(\mu(A_\iota))_{\iota\in J}$ is summable there exists $\iota\in J$ with $\mu(A_\iota) \in U$ and this is a contradiction.

b) follows immediately from a).

c) Let U be a 0-neighbourhood in E . Assume for any finite subset I of $\mathbb{N}$ there exists a finite subset J of $\mathbb{N} \setminus I$ such that

$$\sum_{n\in J} (\mu(A_{n+1}) - \mu(A_n)) \notin U.$$

We may construct inductively a disjoint sequence $(I_m)_{m\in\mathbb{N}}$ of finite subsets of $\mathbb{N}$ such that

$$\sum_{n\in I_m} (\mu(A_{n+1}) - \mu(A_n)) \notin U$$

for any $m\in\mathbb{N}$. Then $(\bigcup_{n\in I_m} (A_{n+1} \setminus A_n))_{m\in\mathbb{N}}$ is a disjoint sequence in $\underline{\underline{R}}$ and

$$\mu(\bigcup_{n\in I_m} (A_{n+1} \setminus A_n)) = \sum_{n\in I_m} (\mu(A_{n+1}) - \mu(A_n)) \notin U$$

for any $m\in\mathbb{N}$. By a) this is a contradiction.

d) Let λ be a positive real measure on $\underline{\underline{R}}$. If λ is not bounded then there exists an increasing sequence $(A_n)_{n\in\mathbb{N}}$ in $\underline{\underline{R}}$ with

$$\lim_{n\to\infty} \lambda(A_n) = \infty .$$

By c) $\lambda((A_n))_{n\in\mathbb{N}}$ is a Σ-sequence and therefore a convergent sequence which is a contradiction. Hence λ is bounded. From these considerations we deduce immediately that μ is bounded. □

Proposition 4.2.8 *Let $\mu\in M(E)$. If $\underline{\underline{R}}$ is a quasi-σ-ring and if the one point sets of E are G_δ-sets then :*

a) any disjoint family $(A_\iota)_{\iota\in I}$ in $\underline{\underline{R}}$ is countable if $\mu(A_\iota) \neq 0$ for any $\iota\in I$;

b) there exists an increasing sequence $(A_n)_{n\in\mathbb{N}}$ in $\underline{\underline{R}}$ such that $\mu(A) = 0$ for any $A\in\underline{\underline{R}}$ with $A\cap(\bigcup_{n\in\mathbb{N}} A_n) = \emptyset$;

c) for any $A\in\underline{\underline{R}}$ with $\mu(A) \neq 0$ there exists $B\in\underline{\underline{R}}$ and $x'\in E'$ such that

$$B\subset A , \ |x'\circ\mu|(B) \neq 0$$

$$\forall C\in\underline{\underline{R}} , \ C\subset B, \ |x'\circ\mu|(C) = 0 \Longrightarrow \mu(C) = 0 .$$

d) there exists $\lambda\in M_+$ such that $x'\circ\mu\ll\lambda$ for any $x'\in E'$.

a) Let $(U_n)_{n\in\mathbb{N}}$ be a sequence of 0-neighbourhoods in E with

$$\bigcap_{n\in\mathbb{N}} U_n = \{0\} .$$

Then

$$\{\iota\in I \mid \mu(A_\iota)\notin U_n\}$$

is a finite set for any $n\in\mathbb{N}$ (Proposition 4.2.7 a)) and from

$$I = \bigcup_{n\in\mathbb{N}} \{\iota\in I \mid \mu(A_\iota)\notin U_n\}$$

it follows that I is countable.

b) Let us denote by Ω the set of sets $\underline{\underline{S}}$ of pairwise disjoint sets of $\underline{\underline{R}}$ such that $\mu(S) \neq 0$ for any $S \in \underline{\underline{S}}$. It is obvious that Ω is inductively ordered by the inclusion relation. By Zorn's lemma there exists a maximal element $\underline{\underline{S}}_o$ of Ω. By a) $\underline{\underline{S}}_o$ is countable. Hence there exists an increasing sequence $(A_n)_{n \in \mathbb{N}}$ in $\underline{\underline{R}}$ such that

$$\bigcup_{A \in \underline{\underline{S}}_o} A = \bigcup_{n \in \mathbb{N}} A_n .$$

Let $A \in \underline{\underline{R}}$ with $A \cap (\bigcup_{n \in \mathbb{N}} A_n) = \emptyset$. If $\mu(A) \neq 0$ then $\underline{\underline{S}}_o \cup \{A\} \in \Omega$ and this contradicts the maximality of $\underline{\underline{S}}_o$.

c) There exists $x' \in E'$ with $(x' \circ \mu)(A) \neq 0$. We set

$$\underline{\underline{S}} := \{D \in \underline{\underline{R}} \mid \exists D', D'' \in \underline{\underline{R}},\ D \cup D' \subset D'' \subset A,\ \mu(D') \neq 0,\ |x' \circ \mu|(D'') = 0\}.$$

$\underline{\underline{S}}$ is a σ-ring and therefore by b) there exists $D \in \underline{\underline{S}}$ such that $\mu(C) = 0$ for any $C \in \underline{\underline{S}}$ with $C \cap D = \emptyset$. By the definition of $\underline{\underline{S}}$ there exist $D', D'' \in \underline{\underline{R}}$ such that

$$D \cup D' \subset D'' \subset A , \quad \mu(D') \neq 0, \quad |x' \circ \mu|(D'') = 0 .$$

We set $B := A \setminus D''$. Then

$$B \in \underline{\underline{R}} , \quad B \subset A , \quad |x' \circ \mu|(B) \neq 0 .$$

Let $C \in \underline{\underline{R}}$ with $C \subset B$ and $|x' \circ \mu|(C) = 0$. If $\mu(C) \neq 0$ then $C \in \underline{\underline{S}}$. Since $C \cap D = \emptyset$ we get $\mu(C) = 0$ (A simpler proof by Hahn's theorem).

d) Let us denote by Ω the set of subsets Δ of $E' \times \underline{\underline{R}}$ possessing the following properties:

$$(x', A) \in \Delta \Longrightarrow |x' \circ \mu|(A) \neq 0,\ (\forall B \in \underline{\underline{R}}, B \subset A,\ |x' \circ \mu|(B) = 0 \Longrightarrow \mu(B) = 0),$$

$$(x', A) ,\ (y', B) \in \Delta ,\ (x', A) \neq (y', B) \Longrightarrow A \cap B = \emptyset .$$

Ω is obviously inductively ordered by the inclusion relation and therefore by Zorn's lemma it possesses a maximal element Δ_o. By a) Δ_o is countable. Let $\varphi : \Delta_o \rightarrow \mathbb{N}$ be an injection. By Proposition 4.2.7 d)

$x'\circ\mu$ is bounded for any $x'\in E'$. We set

$$\lambda := \sum_{(x',A)\in\Delta_o} \frac{1}{2^{\varphi(x',A)}\,\|x'\circ\mu\|} \, |x'\circ\mu| \; .$$

λ is a positive measure on $\underline{\underline{R}}$.

Let $A\in\underline{\underline{R}}$ with $\lambda(A) = 0$. Assume there exists $x'\in E$ with $|x'\circ\mu|(A) \neq 0$. Then there exists $B\in\underline{\underline{R}}$ with $B\subset A$ and $\mu(B) \neq 0$. Let $(y',C)\in\Delta_o$. Then $|y'\circ\mu|(B\cap C) = 0$ and therefore $\mu(B\cap C) = 0$. We set

$$B_o := \bigcup_{(y',C)\in\Delta_o} (B\cap C) \; .$$

Then $B_o\in\underline{\underline{R}}$ and $\mu(B_o) = 0$ and therefore $\mu(B\setminus B_o) \neq 0$. By c) there exists $C\in\underline{\underline{R}}$ and $y'\in E'$ such that

$$C\subset B\setminus B_o \; , \quad |y'\circ\mu|(C) \neq 0 \; ,$$

$$\forall D\in\underline{\underline{R}} \; , \; D\subset C \; , \; |x'\circ\mu|(D) = 0 \Longrightarrow \mu(D) = 0 \; .$$

Hence $\Delta_o \cup\{(y',C)\}\in\Omega$ and this contradicts the maximality of Δ_o. Hence $|x'\circ\mu|(A) = 0$ for any $x'\in E$. We get $x'\circ\mu \ll \lambda$ for any $x'\in E'$. □

<u>*Proposition 4.2.9*</u> *Let $\mu\in M(E)$ and let $\xi\in\hat{L}^1(\mu)$. Let $\underline{\underline{V}}$ be the set of convex circled absorbing sets of E for which there exists a sequence $(U_n)_{n\in\mathbb{N}}$ of 0-neighbourhoods in E with $\bigcap_{n\in\mathbb{N}} U_n\subset V$, let $\underline{\underline{M}}$ be the set of subspaces M of E' which are unions of countable families of equicontinuous sets of E', and let φ be a map of $\underline{\underline{M}}$ into E such that*

$$< \varphi(M) \; , \; x'> = \int \xi d(x'\circ\mu)$$

for any $M\in\underline{\underline{M}}$ and $x'\in M$. Let us order $\underline{\underline{M}}$ by the inclusion relation, let $\underline{\underline{F}}$ be its section filter, and let $\varphi(\underline{\underline{F}})$ be the image of $\underline{\underline{F}}$ with respect to φ. Then :

a) $\varphi(\underline{\underline{F}})$ is a δ-filter on E and a Cauchy filter with respect to the locally convex topology on E for which $\underline{\underline{V}}$ is a fundamental *system of 0-neighbourhoods ;*

b) if $\varphi(\underline{\underline{F}})$ *converges to an* $x\in E$ *then*

$$\int \xi d\mu = x .$$

a) Since the union of any countable family in $\underline{\underline{M}}$ belongs to $\underline{\underline{M}}$, $\underline{\underline{F}}$ is a δ-filter and the same holds then for $\varphi(\underline{\underline{F}})$.

Let $V\in\underline{\underline{V}}$ and let $(U_n)_{n\in\mathbb{N}}$ be a sequence of closed convex circled 0-neighbourhoods in E with $\bigcap_{n\in\mathbb{N}} U_n \subset V$. Let U_n^{o} denote the polar set of U_n in E' for any $n\in\mathbb{N}$ and let M be the subspace of E' generated by $\bigcup_{n\in\mathbb{N}} U_n^{o}$. Then $M\in\underline{\underline{M}}$. Let $M'\in\underline{\underline{M}}$ with $M\subset M'$. We have

$$< \varphi(M) - \varphi(M') , x'> = <\varphi(M), x'> - <\varphi(M') , x'> = 0$$

for any $x'\in M$. We get

$$\varphi(M) - \varphi(M')\in \bigcap_{n\in\mathbb{N}} U_n \subset V .$$

This shows that $\varphi(\underline{\underline{F}})$ is a Cauchy filter with respect to the locally convex topology on E for which $\underline{\underline{V}}$ is a fundamental system of neighbourhoods.

b) Let $x'\in E'$. Let $M\in\underline{\underline{M}}$ with $x'\in M$. Then

$$\int \xi d(x'\circ\mu) = <\varphi(M') , x'>$$

for any $M'\in\underline{\underline{M}}$ with $M \subset M'$ and therefore

$$\int \xi d(x'\circ\mu) = \lim_{M',\underline{\underline{F}}} <\varphi(M'), x'> = <x, x'> .$$

Since x' is arbitrary we get

$$\int \xi d\mu = x . \square$$

<u>*Theorem 4.2.10*</u> *Let* $\underline{\underline{R}}$ *be a quasi-*σ*-ring, let* $\mu\in M(E)$*, let* $\underline{\underline{V}}_1$ *be the set of convex circled absorbing sets of* E *which absorb* $\mu(\underline{\underline{R}})$ *,*

let $\underline{\underline{V}}_2$ *be the set of* $V\in\underline{\underline{V}}_1$ *for which there exists a sequence* $(U_n)_{n\in\mathbb{N}}$ *of 0-neighbourhoods in* E *with* $\bigcap_{n\in\mathbb{N}} U_n \subset V$, *and for any* $i\in\{1,2\}$ *let* $\underline{\underline{T}}_i$ *be the locally convex topology on* E *for which* $\underline{\underline{V}}_i$ *is a fundamental system of 0-neighbourhoods. If the* Σ*-sequences in* $(E,\underline{\underline{T}}_1)$ *and the bounded Cauchy* δ*-filters on* $(E,\underline{\underline{T}}_2)$ *are convergent in* E *then* $\xi\in\hat{L}^1(\mu)$ *and* $\int\xi d\mu\in E$ *for any* $\xi\in M_c^\pi$.

Let F be an arbitrary subspace of E'. We denote by E(F) the dual of F_E endowed with the topology of uniform convergence on the equicontinuous sets of F. For any $x\in E$ we denote by x_F the element of E(F) defined by

$$F \longrightarrow \mathbb{R}\ ,\ x' \longmapsto \langle x,x'\rangle\ .$$

We set

$$\mu_F : \underline{\underline{R}} \longrightarrow E(F)\ ,\ A \longmapsto \mu(A)_F\ .$$

Then $\mu_F\in M(E(F))$.

We denote by $\underline{\underline{M}}$ the set of subspaces M of E' which are unions of countable families of equicontinuous sets of E'. We order $\underline{\underline{M}}$ by the inclusion relation and denote by $\underline{\underline{F}}$ the section filter on $\underline{\underline{M}}$. We denote further by $\underline{\underline{S}}$ the set of $\underline{\underline{R}}$-measurable subsets of X.

Let $A\in\underline{\underline{S}}$ and let $M\in\underline{\underline{M}}$. Since the one point sets of E(M) are G_δ-sets there exists by Proposition 4.2.8 b) an increasing sequence $(A_n)_{n\in\mathbb{N}}$ in $\underline{\underline{R}}$ such that $\mu_M(B) = 0$ for any $B\in\underline{\underline{R}}$ with $B\cap(\bigcup_{n\in\mathbb{N}} A_n)=\emptyset$. By Proposition 4.2.7 c) $(\mu(A\cap A_n))_{n\in\mathbb{N}}$ is a Σ-sequence and therefore a convergent sequence in E. We set

$$x_M := \lim_{n\to\infty} \mu(A\cap A_n)\ .$$

For any $x'\in M$ the set $A\setminus\bigcup_{n\in\mathbb{N}} A_n$ is an $x'\circ\mu$-null set and since $x'\circ\mu$ is bounded (Proposition 4.2.7 d)) we have $1_A\in L^1(x'\circ\mu)$. We get

$$\int 1_A d(x'\circ\mu) = \lim_{n\to\infty} (x'\circ\mu)(A\cap A_n) = \langle x_M,x'\rangle\ .$$

By Proposition 4.2.9 a) the image of $\underline{\underline{F}}$ with respect to the map

$$\underline{\underline{M}} \longrightarrow E \ , \ M \longmapsto x_M$$

is a Cauchy δ-filter on $(E,\underline{\underline{T}}_2)$. Since it is also a bounded filter on $(E,\underline{\underline{T}}_2)$ it converges to an $x \in E$. By Proposition 4.2.9 b)

$$\int 1_A d\mu = x \in E \ .$$

Let f be a bounded $\underline{\underline{R}}$-measurable function on X. There exists a sequence $(f_n)_{n\in\mathbb{N}}$ of step functions on X with respect to $\underline{\underline{S}}$ such that

$$|f-f_n| < \frac{1}{2^n}$$

for any $n\in\mathbb{N}$. By the above proof $\int f_n d\mu \in E$ for any $n\in\mathbb{N}$. $(\dot{f}_n)_{n\in\mathbb{N}}$ is a Σ-sequence in M_b^π and therefore, by Proposition 4.2.5 d), $(\int f_n d\mu)_{n\in\mathbb{N}}$ is a Σ-sequence in E with respect to $\underline{\underline{T}}_1$. Hence $(\int f_n d\mu)_{n\in\mathbb{N}}$ converges in E to an x. We have

$$\langle x,x'\rangle = \lim_{n\to\infty} \langle \int f_n d\mu, x'\rangle = \lim_{n\to\infty} \int f_n d(x' \circ \mu) = \int f d(x' \circ \mu)$$

for any $x' \in E'$ and therefore

$$\int f d\mu = x \in E \ .$$

Let $\xi \in M_c$. By Proposition 4.2.7 d) $M = M_b$ and therefore, by Proposition 3.3.3, $\xi \in M_b$. We get $\xi \in \hat{L}^1(\mu)$. Let $M \in \underline{\underline{M}}$. Since the one point sets of $E(M)$ are G_δ-sets there exists by Proposition 4.2.8 d) a $\lambda \in M$ such that $x' \circ \mu \ll \lambda$ for any $x' \in M$. By Theorem 3.2.4 f) there exists a bounded $\underline{\underline{R}}$-measurable real function f_M on X such that the components of $\dot{f}_M$ and ξ on $L^\perp(\lambda)$ coincide. We may even assume $\|\dot{f}_M\| \leq \|\xi\|$, where $\|\cdot\|$ denotes the norm of M_b^π. We have

$$\int f_M d(x' \circ \mu) = \int \xi d(x' \circ \mu)$$

for any $x' \in M$. By the above proof $\int f_M d\mu \in E$ and by Proposition 4.2.5

$$\int f_M d\mu \in 2\|\xi\| \Gamma\mu(\underline{\underline{R}}) \ ,$$

where $\Gamma\mu(\underline{\underline{R}})$ denotes the closed convex circled hull of $\mu(\underline{\underline{R}})$ in $E''_{E'}$.

Hence the image of $\underline{\underline{F}}$ with respect to the map

$$\underline{\underline{M}} \longrightarrow E \, , \, M \longmapsto \int f_M d\mu$$

is a bounded filter on $(E,\underline{\underline{T}}_2)$. By Proposition 4.2.9 a) it is a Cauchy δ-filter on $(E,\underline{\underline{T}}_2)$. Hence it converges to an $x \in E$. By Proposition 4.2.9 b)

$$\int \xi d\mu = x \in E \; . \; \Box$$

Remarks.

1. Let λ be the Lebesgue measure on $\mathbb{R}$, let $\underline{\underline{R}}$ be the set of Borel subsets of $\mathbb{R}$ such that

$$\lim_{n\to\infty} \lambda(A \cap [n,n+1[) = 0 \, ,$$

and let M be the set of real measures on $\underline{\underline{R}}$. Further let $\mu \in M \setminus M_b$. Then there exists a sequence $(m(n))_{n\in\mathbb{N}}$ in $\mathbb{N}\setminus\{0\}$ such that

$$\lim_{n\to\infty} m(n) = \infty, \quad \sum_{n\in\mathbb{N}} \frac{1}{m(n)} |\mu|([n,n+1[) = \infty \, .$$

For any $n \in \mathbb{N}$ there exists $i(n) \in \mathbb{N}$ such that $1 \leqslant i(n) \leqslant m(n)$ and

$$|\mu|([n+\frac{i(n)-1}{m(n)} \; , \; n+\frac{i(n)}{m(n)}[) \geqslant \frac{1}{m(n)} |\mu|([n,n+1[) \; .$$

We get

$$A := \bigcup_{n\in\mathbb{N}} [n+\frac{i(n)-1}{m(n)} \; , \; n+\frac{i(n)}{m(n)}[\in \underline{\underline{R}}$$

and this leads to the contradictory relation

$$|\mu|(A) = \sum_{n\in\mathbb{N}} |\mu|([n+\frac{i(n)-1}{m(n)} \; , \; n+\frac{i(n)}{m(n)}[) = \infty \; .$$

Hence $M = M_b$.

Let E be the Banach space c_o ($c_o := \{(\alpha_n)_{n\in\mathbb{N}} \in \mathbb{R}^{\mathbb{N}} \mid \lim_{n\to\infty} \alpha_n = 0\}$

with norm

$$\|(\alpha_n)_{n\in\mathbb{N}}\| := \sup_{n\in\mathbb{N}} |\alpha_n|) .$$

We denote for any $A\in\underline{\underline{R}}$ by $\mu(A)$ the map

$$\mathbb{N} \longrightarrow \mathbb{R} , \ n \longmapsto \lambda(A\cap[n,n+1[) .$$

Then $\mu(A)\in E$ for any $A\in\underline{\underline{R}}$ and $\mu\in M(E)$ but $\int 1 d\mu \notin E$. Hence we cannot replace the hypothesis "$\underline{\underline{R}}$ is a quasi-σ-ring" in the above result by the weaker one "$M = M_b$".

2. Let E be c_o, let $\underline{\underline{R}}$ be the set of subsets of N, and let M be the set of real measures on $\underline{\underline{R}}$. We set

$$f : N \longrightarrow \mathbb{R} , \ n \longmapsto \frac{1}{2^n} ,$$

$$\mu : \underline{\underline{R}} \longrightarrow E , \ A \longmapsto f 1_A ,$$

$$N := \{\lambda\in M \,|\, \{n\in\mathbb{N} \,|\, \lambda(\{n\}) \neq 0\} \text{ is finite } \} ,$$

$$\xi : N \longrightarrow \mathbb{R} , \ \lambda \longmapsto \sum_{n\in\mathbb{N}} 2^n \lambda(\{n\}) .$$

Then $\mu\in M(E)$, $N\in\Phi$, $\xi\in N^\pi \subset M^\rho$, $\xi\in\hat{L}^1(\mu)$, and $\int\xi d\mu\notin E$. This example shows that we cannot replace the hypothesis "$\xi\in M_c^\pi$" in the above result with "$\xi\in\hat{L}^1(\mu)$" .

3. Let X be an uncountable set, let $\mathbb{R}^X$ be the vector space of real functions on X endowed with the locally convex topology of pointwise convergence (i.e. with the product topology), let E be its subspace

$$\{f\in\mathbb{R}^X \,|\, \{f \neq 0\} \text{ is countable}\} ,$$

let $\underline{\underline{R}}$ be the σ-ring of countable subsets of X, let M be the set of real measures on $\underline{\underline{R}}$, and let μ be the map

$$\underline{\underline{R}} \longrightarrow E , \ A \longmapsto 1_A .$$

Then $\mu \in M_b(E)$, E is sequentially complete, but the condition concerning $(E,\underline{\underline{T}}_2)$ in the above result is not fulfilled. We have $\int 1 d\mu \notin E$. Hence we cannot drop the hypothesis concerning $(E,\underline{\underline{T}}_2)$ in the above result.

4. Let X be $\mathbb{N}$, let $\underline{\underline{R}}$ be the power set of $\mathbb{N}$, let M be the set of real measures on $\underline{\underline{R}}$, let E be the vector space of real functions on $\mathbb{N}$ taking a finite number of values only endowed with the topology of pointwise convergence, and let μ be the map

$$\underline{\underline{R}} \longrightarrow E \ , \ A \longmapsto 1_A \ .$$

Then $\mu \in M_b(E)$, E is δ-complete (since it is metrizable), but the hypothesis concerning $(E,\underline{\underline{T}}_1)$ in the theorem is not fulfilled. If we set

$$f : X \longrightarrow \mathbb{R} \ , \ n \longmapsto \frac{1}{n}$$

then $\int f d\mu \notin E$. This example shows that we cannot drop the condition concerning $(E,\underline{\underline{T}}_1)$ in the theorem.

<u>Theorem 4.2.11</u> *If $\mu \in M_b(E)$ then the following assertions are equivalent :*

a) $\int \xi d\mu \in E$ for any $\xi \in M_b^\pi$;

b) the closed convex circled hull of $\mu(\underline{\underline{R}})$ is weakly compact.

If these conditions are fulfilled then :

c) the set

$$\{\int \xi d\mu \mid \xi \in M_b^\pi \ , \ \|\xi\| \leq 1\}$$

is weakly compact and it is compact if $\mu(\underline{\underline{R}})$ is precompact ;

d) the map

$$M_b^\pi \longrightarrow E \ , \ \xi \longmapsto \int \xi d\mu$$

is continuous and order continuous ;

e) the set $\{\int \xi d\mu \mid \xi \in F\}$ is precompact for any weakly pseudocompact set F of M_b^π ;

f) let F be a subset of M_{b+}^π such that $\bigvee_{\eta \in F} \eta = 1$ and for any

$\eta\in F$ *let* r_η *be the seminorm*

$$M_b^\pi \longrightarrow \mathbb{R}_+ \ , \ \xi \longmapsto \|\xi\eta\| ;$$

if we endow M_b^π *with the locally convex topology generated by the family* $(r_\eta)_{\eta\in F}$ *of seminorms then the restriction of the map*

$$M_b^\pi \longrightarrow E \ , \ \xi \longmapsto \int \xi d\mu$$

to the set $\{\xi\in M_b^\pi \mid \|\xi\| \leqslant 1\}$ *is uniformly continuous.*

All the above assertions a) - f) *are fulfilled if* $\underline{\underline{R}}$ *is a quasi-*σ*-ring and if* E *is* Σ*-complete and* δ*-quasicomplete (e.g.* E *quasicomplete).*

a $\Longrightarrow$ b. The map

$$M_b^\pi \longrightarrow E \ , \ \xi \longmapsto \int \xi d\mu$$

is the adjoint of the map

$$E' \longrightarrow M_b \ , \ x' \longmapsto x' \circ \mu$$

and therefore continuous for the corresponding weak topologies. The set $\{\xi\in M_b^\pi \mid \|\xi\| \leqslant 1\}$ being compact for the weak topology $\sigma(M_b^\pi , M_b)$ associated to the duality $\langle M_b^\pi , M_b\rangle$, its image

$$\{\int \xi d\mu \mid \xi\in M_b^\pi , \|\xi\| \leqslant 1\}$$

is weakly compact. This set is closed convex circled and contains $\mu(\underline{\underline{R}})$ and therefore the closed convex circled hull of $\mu(\underline{\underline{R}})$ is weakly compact too.

b $\Longrightarrow$ a & c. The closed convex circled hull of $\mu(\underline{\underline{R}})$ in E and in $E''_{E'}$ coincide. By Proposition 4.2.5

$$\{\int \xi d\mu \mid \xi\in M_b^\pi , \|\xi\| \leqslant 1\}$$

is a weakly compact set of E . We get immediately $\int \xi d\mu \in E$ for any $\xi\in M_b^\pi$. If $\mu(\underline{\underline{R}})$ is precompact then its closed convex circled hull is compact and by Proposition 4.2.5 the set

$$\{\int \xi d\mu \mid, \ \xi \in M_b^\pi \|\xi\| \leq 1\}$$

is compact too.

a $\Longrightarrow$ d . By a $\Longrightarrow$ c the map

$$M_b^\pi \longrightarrow E \ , \ \xi \longmapsto \int \xi d\mu$$

is continuous. By Proposition 4.1.6 it is order continuous.

d $\Longrightarrow$ e follows immediately from Proposition 3.8.5 c $\Longrightarrow$ a .

d $\Longrightarrow$ f . We may assume F upper directed. Let V be a closed convex circled 0-neighbourhood in E . Assume for any $\eta \in F$ and for any $\varepsilon > 0$ there exists $\xi \in M_b^\pi$ such that

$$\|\xi\| \leq 1 \ , \ r_\eta(\xi) < \varepsilon \ , \ \int \xi d\mu \notin V \ .$$

Then for any $\eta \in F$ and for any $\varepsilon > 0$ there exists $\xi \in M_b^\pi$ such that

$$0 \leq \xi \leq \dot{1} \ , \ r_\eta(\xi) < \varepsilon \ , \ \int \xi d\mu \notin \frac{1}{2} V \ .$$

We construct inductively an increasing sequence $(\eta_n)_{n \in \mathbb{N}}$ in F and a sequence $(\xi_n)_{n \in \mathbb{N}}$ in M_b^π such that

$$0 \leq \xi_n \leq \dot{1} \ , \ r_{\eta_{n-1}}(\xi_n) < \frac{1}{n+1} \quad \int \xi_n \eta_n d\mu \notin \frac{1}{2} V$$

for any $n \in \mathbb{N}$, where $\eta_{-1} = 0$. Let $n \in \mathbb{N}$ and assume the sequences were constructed up to $n - 1$. There exists $\xi_n \in M_b^\pi$ such that

$$0 \leq \xi_n \leq \dot{1} \ , \ r_{\eta_{n-1}}(\xi_n) < \frac{1}{n+1} \quad \int \xi_n d\mu \notin \frac{1}{2} V.$$

$(\xi_n \eta)_{\eta \in F}$ is an upper directed family in M_b^π with supremum ξ_n (Theorem 3.1.7 a)). By d) the net

$$\left(\int \xi_n \eta d\mu\right)_{\eta \in F}$$

converges to $\int \xi_n d\mu$. Hence there exists $\eta_n \in F$ such that

$$\eta_{n-1} \leq \eta_n \ , \int \xi_n \eta_n d\mu \notin \frac{1}{2} V \ .$$

This finishes the inductive construction.

We have $\|\xi_n \eta_{n-1}\| < \frac{1}{n+1}$ for any $n \in \mathbb{N}$. By b) there exists $n_o \in \mathbb{N}$ such that

$$\int \xi_n \eta_{n-1} d\mu \in \frac{1}{4} V$$

for any $n \geq n_o$. We have

$$0 \leq \xi_n (\eta_n - \eta_{n-1}) \leq \eta_n - \eta_{n-1}$$

and therefore $(\sum_{n=0}^{m} \xi_n (\eta_n - \eta_{n-1}))_{m \in \mathbb{N}}$ is an increasing upper bounded sequence in M_b^π . By b) the sequence

$$(\sum_{n=0}^{m} \int \xi_n (\eta_n - \eta_{n-1}) d\mu)_{m \in \mathbb{N}}$$

converges. Hence there exists $n \in \mathbb{N}$ with $n \geq n_o$ and

$$\int \xi_n (\eta_n - \eta_{n-1}) d\mu \in \frac{1}{4} V \ ,$$

and we get the contradictory relation

$$\int \xi_n \eta_n d\mu \in \frac{1}{2} V \ .$$

We deduce that there exists $\eta \in F$ and $\varepsilon > 0$ such that $\int \xi d\mu \in V$ for any $\xi \in M_b^\pi$ with $r_\eta(\xi) < \varepsilon$ and $\|\xi\| \leq 1$. Let $\xi', \xi'' \in M_b^\pi$ with $\|\xi'\| \leq 1$, $\|\xi''\| \leq 1$, and $r_n \ (\xi' - \xi'') < \varepsilon$. Then

$$\int \frac{1}{2} (\xi' - \xi'') d\mu \in V$$

and therefore

$$\int \xi' d\mu - \int \xi'' d\mu \in 2V \ .$$

Assume now $\underline{\underline{R}}$ is a quasi-σ-ring and E is Σ-complete and δ-complete.

Let us denote by $\underline{\underline{T}}_1$ and $\underline{\underline{T}}_2$ the topologies introduced in Theorem 4.2.10. Let $(x_n)_{n\in\mathbb{N}}$ be a Σ-sequence in E with respect to $\underline{\underline{T}}_1$. Since the topology of E is coarser than $\underline{\underline{T}}_1$, $(x_n)_{n\in\mathbb{N}}$ is a Σ-sequence and therefore a convergent sequence in E.

Let now $\underline{\underline{F}}$ be a bounded Cauchy δ-filter on $(E,\underline{\underline{T}}_2)$. Since the topology of E is coarser than $\underline{\underline{T}}_2$, $\underline{\underline{F}}$ is a bounded Cauchy δ-filter and therefore a convergent filter on E.

a) follows now immediately from Theorem 4.2.10. □

Remarks. 1. I. Tweedle showed ([31] Theorem 3) that the convex hull of $\mu(\underline{\underline{R}})$ is weakly relatively compact if $\underline{\underline{R}}$ is a σ-ring and E quasi-complete.

The assertion a) was proved in [5] (Theorem 10) under the hypothesis $\underline{\underline{R}}$ is a σ-ring and E is sequentially complete and δ-complete.

2. $\mu(\underline{\underline{R}})$ is not always relatively compact even if E is a Banach lattice and $\underline{\underline{R}}$ a σ-algebra. Let X be the closed interval $[0,1]$, let $\underline{\underline{R}}$ be the σ-algebra of Borel sets of X, let λ be the Lebesgue measure on $\underline{\underline{R}}$, and let E be the Banach lattice $L^1(\lambda)$. We denote by μ the map

$$\underline{\underline{R}} \longrightarrow E, \quad A \longmapsto \dot{1}_A ,$$

where $\dot{1}_A$ denotes the equivalence class of 1_A in $L^1(\lambda)$. μ is an E-valued measure on $\underline{\underline{R}}$ and $\mu(\underline{\underline{R}})$ is closed but not compact.

Lemma 4.2.12 *Let* $\mathbb{N}$ *be endowed with the discrete topology and let* $\beta\mathbb{N}$ *be the Stone-Čech compactification of* $\mathbb{N}$. *Then:*

a) if $(U_n)_{n\in\mathbb{N}}$ *is a disjoint sequence of open closed nonempty sets of* $\beta\mathbb{N}\setminus\mathbb{N}$ *then* $\bigcup_{n\in\mathbb{N}} U_n$ *is not a dense set of* $\beta\mathbb{N}\setminus\mathbb{N}$;

b) there exists an uncountable disjoint family of open closed nonempty sets of $\beta\mathbb{N}\setminus\mathbb{N}$;

c) $\beta\mathbb{N}\setminus\mathbb{N}$ *is not separable.*

a) We construct inductively a disjoint sequence $(A_n)_{n\in\mathbb{N}}$ of subsets of $\mathbb{N}$ such that the closure of A_n in $\beta\mathbb{N}$ contains U_n and does not meet $\overline{\bigcup_{m\in\mathbb{N}\setminus\{n\}} U_m}$ for any $n\in\mathbb{N}$.

Let $n\in\mathbb{N}$ and assume the sequence was constructed up to $n-1$. U_n and $\overline{\bigcup_{m\in\mathbb{N}\setminus\{n\}} U_m}$ are closed disjoint sets of $\beta\mathbb{N}$ and therefore there exists an open and closed set U of $\beta\mathbb{N}$ containing U_n and not meeting $\overline{\bigcup_{m\in\mathbb{N}\setminus\{n\}} U_m}$

We set

$$A_n := (\mathbb{N}\setminus\bigcup_{m<n} A_m)\cap U .$$

We have

$$U_n \subset U \subset \overline{\bigcup_{m\leq n} A_m} = \bigcup_{m\leq n} \bar{A}_m .$$

Since $U_n \cap \bar{A}_m = \emptyset$ for any $m<n$ we get $U_n \subset \bar{A}_n$. This finishes the construction. Let x_n be a point of A_n for any $n\in\mathbb{N}$. Then $\overline{\{x_n \mid n\in\mathbb{N}\}}\setminus\mathbb{N}$ is a nonempty open set of $\beta\mathbb{N}\setminus\mathbb{N}$ which does not meet $\bigcup_{m\in\mathbb{N}} U_n$.

b) Let Ω be the set of infinite sets $\underline{\underline{U}}$ of pairwise disjoint open and closed sets of $\beta\mathbb{N}\setminus\mathbb{N}$. It is obvious that Ω is inductively ordered by the inclusion relation and therefore by Zorn's lemma it possesses a maximal element $\underline{\underline{U}}_0$. The set $\bigcup_{U\in\underline{\underline{U}}_0} U$ is then a dense set of $\beta\mathbb{N}\setminus\mathbb{N}$ and by a) $\underline{\underline{U}}_0$ is not countable.

c) follows immediately from b). □

Corollary 4.2.13 *Let* I *be a set, let* $\underline{\underline{J}}$ *be a quasi-σ-ring of subsets of* I *containing any finite subset of* I, *let* $\ell^\infty(I)$ *be the vector lattice*

$$\{(\alpha_\iota)_{\iota\in I}\in\mathbb{R}^I \mid \sup_{\iota\in I} |\alpha_\iota| < \infty\}$$

endowed with the norm

$$\ell^\infty(I) \longrightarrow \mathbb{R}_+ , \quad (\alpha_\iota)_{\iota\in I} \longmapsto \sup_{\iota\in I}(\alpha_\iota)$$

and let $(x_\iota)_{\iota\in I}$ *be a family in* E *such that* $(x_\iota)_{\iota\in J}$ *is summable for any* $J\in\underline{\underline{J}}$. *If* E *is* Σ-*complete and* δ-*quasicomplete (e.g. quasicomplete) Then:*

a) $(\alpha_\iota x_\iota)_{\iota\in I}$ *is summable for any* $(\alpha_\iota)_{\iota\in I}\in\ell^\infty(I)$;

b) *the map*

$$\ell^\infty(I) \longrightarrow E\ ,\ (\alpha_\iota) \longmapsto \sum_{\iota\in I}\alpha_\iota x_\iota\ .$$

is continuous and order continuous ;

c) *the map*

$$\{(\alpha_\iota)_{\iota\in I}\in\ell^\infty(I) \mid \sup_{\iota\in I}|\alpha_\iota| \leqslant 1\} \longrightarrow E\ ,\ (\alpha_\iota)_{\iota\in I} \longmapsto \sum_{\iota\in I}\alpha_\iota x_\iota\ .$$

is uniformly continuous with respect to the uniformity of pointwise convergence on $\{(\alpha_\iota)_{\iota\in I}\in\ell^\infty(I) \mid \sup_{\iota\in I}|\alpha_\iota| \leqslant 1\}$ *(i.e. the uniformity induced by the product uniformity on* $\mathbb{R}^I$) ;

d) *the set*

$$\{\sum_{\iota\in I}\alpha_\iota x_\iota \mid (\alpha_\iota)_{\iota\in I}\in\ell^\infty(I)\ ,\ \sup_{\iota\in I}|\alpha_\iota| \leqslant 1\}$$

is compact.

As $\underline{\underline{J}}$ *we may take*

$$\{J\subset I \mid J \text{ countable}\ ,\ J\notin \bigcup_{\underline{\underline{F}}\in\Psi}\underline{\underline{F}}\}\ ,$$

where Ψ *is a countable set of free ultrafilters on* I .

We set

$$X := I\ ,\ \underline{\underline{R}} := \underline{\underline{J}}$$

$$M := \{\lambda \mid \lambda \text{ real measure on } \underline{\underline{R}}\}\ ,$$

$$\mu : \underline{\underline{R}} \longrightarrow E\ ,\ J \longmapsto \sum_{\iota\in J} x_\iota\ .$$

$\underline{\underline{R}}$ is a quasi-σ-ring and $\mu\in M(E)$. It is easy to see that $\ell^\infty(I)$ may be identified with M_b^π as Banach lattices.

a) follows from the last remark of Theorem 4.2.11.

b) follows from a) and Theorem 4.2.11 a $\Longrightarrow$ d .

c) We set

$$F := \{1_J \mid J \subset I \text{ , } J \text{ finite}\} .$$

Then F is a subset of M_{b+}^π such that $\bigvee_{\eta \in F} \eta = \dot{1}$. For any $\eta \in F$ let r_η be the seminorm

$$M_b^\pi \longrightarrow \mathbb{R}_+ \text{ , } \xi \longmapsto \|\xi\eta\| .$$

Then the locally convex topology on M_b^π generated by the family $(r_\eta)_{\eta\in F}$ of seminorms is nothing else but the topology of pointwise convergence and the assertion follows from a) and Theorem 4.2.11 a $\Longrightarrow$ f.

d) The set $\{(\alpha_\iota)_{\iota\in I} \in \ell^\infty(I) \mid \sup_{\iota\in I}|\alpha_\iota| \leq 1\}$ is compact with respect to the topology of pointwise convergence and the assertion follows from c).

Let now Ψ be a countable set of free ultrafilters on I and assume

$$\underline{\underline{J}} := \{J \subset I \mid J \text{ countable, } J \notin \bigcup_{\underline{\underline{F}} \in \Psi} \underline{\underline{F}}\} .$$

It is obvious that $\underline{\underline{J}}$ is a δ-ring. Let $(J_n)_{n\in\mathbb{N}}$ be a disjoint sequence in $\underline{\underline{J}}$. Let φ be the map

$$I \longrightarrow \mathbb{N} \text{ , } \iota \longmapsto \begin{cases} n & \text{if } \iota \in J_n \\ 0 & \text{if } \iota \notin \bigcup_{n\in\mathbb{N}} J_n \end{cases} .$$

We set

$$\Psi' := \{\varphi(\underline{\underline{F}}) \mid \underline{\underline{F}} \in \Psi\} .$$

Let $\underline{\underline{U}}$ be an uncountable set of pairwise disjoint open nonempty sets of $\beta\mathbb{N}\setminus\mathbb{N}$ (Lemma 4.2.12 b)). Since Ψ' is countable there exists $U \in \underline{\underline{U}}$ such that each $\underline{\underline{F}}'\in\Psi'$ converges to a point of $\beta\mathbb{N}\setminus U$. Let M be an infinite subset of $\mathbb{N}\setminus\{0\}$ such that $\bar{M}\cap\beta\mathbb{N}\subset U$. Then $M\notin\bigcup_{\underline{\underline{F}}'\in\Psi'}\underline{\underline{F}}'$. We get

$$\bigcup_{n\in M} J_n = \overset{-1}{\varphi}(M)\notin\bigcup_{\underline{\underline{F}}\in\Psi}\underline{\underline{F}}\ .$$

Hence $\underline{\underline{J}}$ is a quasi-σ-ring. □

<u>*Corollary 4.2.14*</u> *Let I be a set, let $\ell^1(I)$, $\ell^\infty(I)$ be the vector spaces*

$$\{(\alpha_\iota)_{\iota\in I}\ \mathbb{R}^I \mid \sum_{\iota\in I}|\alpha_\iota| < \infty\}\ ,$$

$$\{(\alpha_\iota)_{\iota\in I}\ \mathbb{R}^I \mid \sup_{\iota\in I}|\alpha_\iota| < \infty\}$$

respectively and let $<\ ,\ >$ be the duality

$$\ell^1(I)\times\ell^\infty(I)\longrightarrow\mathbb{R}\ ,((\alpha_\iota)_{\iota\in I}\ ,\ (\beta_\iota)_{\iota\in I})\longmapsto\sum_{\iota\in I}\alpha_\iota\beta_\iota\ .$$

Then

$$\{(\alpha_\iota)_{\iota\in I}\in\ell^\infty(I)\mid\sup_{\iota\in I}|\alpha_\iota|\leqslant 1\}$$

is compact with respect to the Mackey topology on $\ell^\infty(I)$ associated to the above duality.

Let us denote by E the vector space $\ell^\infty(I)$ endowed with the Mackey topology associated to the duality $<\ell^\infty(I)\ ,\ \ell^1(I)>$. For any $\iota\in I$ we denote by x_ι the element of E

$$I\longrightarrow\mathbb{R}\ ,\ \iota'\longmapsto\begin{cases}1 & \text{if } \iota'=\iota\\ 0 & \text{if } \iota'\neq\iota\ .\end{cases}$$

Then $(x_\iota)_{\iota\in J}$ is summable in E for any $J\subset I$. By Corollary 4.2.13 d)

$$\{(\alpha_\iota)_{\iota\in I}\in\ell^\infty(I)\mid\sup_{\iota\in I}|\alpha_\iota|\leqslant 1\} = \{\sum_{\iota\in I}\alpha_\iota x_\iota\mid(\alpha_\iota)_{\iota\in I}\in\ell^\infty(I),\ \sup_{\iota\in I}|\alpha_\iota|\leqslant 1\}$$

is a compact set of E . □

Remark. In the above corollary we have an example of a measure λ such that the unit ball of $L^{\infty}(\lambda)$ is compact with respect to the Mackey topology on $L^{\infty}(\lambda)$ associated to the duality $\langle L^{\infty}(\lambda), L^{1}(\lambda)\rangle$. This is not always the case as it can be seen by taking λ the Lebesgue measure on $[0,1]$.

Proposition 4.2.15 *Let* $\underline{\underline{R}}^{o}$ *be a* δ*-ring contained in* $\underline{\underline{R}}$ *and let* N *be a solid subspace of* M *containing* M_c *. We set*

$$N^{o} := \{\lambda|\underline{\underline{R}}^{o} \mid \lambda\in N\}, \quad \xi^{o} := \bigvee_{A\in\underline{\underline{R}}^{o}} i_A .$$

If $(\forall A\in\underline{\underline{R}}, \forall A^{o}\in R^{o}, \ A\subset A^{o} \Longrightarrow A\in\underline{\underline{R}}^{o})$ *then:*

a) $1_A\in L^{1}(\lambda^{o})$ *for any* $\lambda^{o}\in N^{o}$ *and* $A\in\underline{\underline{R}}$ *; we denote by* $\varphi\lambda^{o}$ *the map*

$$\underline{\underline{R}} \longrightarrow \mathbb{R}, \ A \longrightarrow \int 1_A d\lambda^{o} ;$$

b) $\varphi\lambda^{o}\in N$ *for any* $\lambda^{o}\in N^{o}$ *; we denote by* P *the solid subspace of* M *generated by* $M_c \cup\{\varphi\lambda^{o} \mid \lambda^{o}\in N^{o}\}$ *;*

c) *for any* $\xi\in P^{\pi}$ *the map*

$$N^{o} \longrightarrow \mathbb{R}, \ \lambda^{o} \longmapsto \xi(\varphi\lambda^{o})$$

belongs to $N^{o\pi}$ *; we denote it by* $\varphi'\xi$ *;*

d) $\xi\xi^{o}\in N^{\pi}$ *for any* $\xi\in P^{\pi}$ *and*

$$\int \xi\xi^{o}\, d\mu = \int \varphi'\xi\, d(\mu|\underline{\underline{R}}^{o})$$

for any $\mu\in N(E)$ *.*

a) Let $\lambda\in N$ with $\lambda|\underline{\underline{R}}^{o} = \lambda^{o}$. We may assume λ (and therefore λ^{o}) positive. 1_A is obviously $\underline{\underline{R}}^{o}$-measurable and

$$\int^{*} 1_A d\lambda^{o} \leq \int 1_A d\lambda < \infty .$$

Hence $1_A\in L^{1}(\lambda^{o})$.

b) Let $\lambda\in N$ with $\lambda|\underline{\underline{R}}^{o} = \lambda^{o}$. We have

$$|\varphi\lambda^{o}(A)| \leqslant |\lambda|(A)$$

for any $A \in \underline{\underline{R}}$. We deduce that $\varphi\lambda^{o}$ is a measure and $|\varphi\lambda^{o}| \leqslant |\lambda|$. Hence $\varphi\lambda^{o} \in N$.

c) is easy to prove.

d) We may assume ξ positive. Let $\lambda \in N_{+}$ and $A \in \underline{\underline{R}}^{o}$. By Proposition 3.2.8 there exists a measurable real function f on X such that

$$f = f1_{A}, \; \dot{f}\cdot\lambda = (\xi\dot{1}_{A})\cdot\lambda, \; f\cdot\varphi(\lambda|\underline{\underline{R}}^{o}) = (\xi\dot{1}_{A})\cdot\varphi(\lambda|\underline{\underline{R}}^{o}) \quad .$$

We get (Theorem 3.2.2 d))

$$\int\xi\dot{1}_{A}d\lambda = \int f d\lambda = \int f d(\varphi(\lambda|\underline{\underline{R}}^{o})) = \int\xi\dot{1}_{A}d(\varphi(\lambda|\underline{\underline{R}}^{o})) \; .$$

We have $\xi\xi^{o} \in P^{\pi}$ and

$$\int\xi\xi^{o}d(\varphi(\lambda|\underline{\underline{R}}^{o})) = \sup_{A\in\underline{\underline{R}}^{o}} \int\xi\dot{1}_{A}d(\varphi(\lambda|\underline{\underline{R}}^{o})) = \sup_{A\in\underline{\underline{R}}^{o}} \int\xi\dot{1}_{A}d\lambda$$

and therefore $\xi\xi^{o} \in L^{1}(\lambda)$ and

$$\int\xi\xi^{o}d\lambda = \int\xi\xi^{o}d(\varphi(\lambda|\underline{\underline{R}}^{o})) \; .$$

Since $\xi(1-\xi^{o}) \in P^{\pi}$ and

$$\int\xi(1-\xi^{o})d(\varphi(\lambda|\underline{\underline{R}}^{o})) = \sup_{A\in\underline{\underline{R}}^{o}}\int\xi(1-\xi^{o})\dot{1}_{A}d(\varphi(\lambda|\underline{\underline{R}}^{o})) = 0$$

we get further

$$\int\xi\xi^{o}d\lambda = \int\xi d(\varphi(\lambda|\underline{\underline{R}}^{o})) \; .$$

λ being arbitrary we deduce $\xi\xi^{o} \in N^{\pi}$ and

$$< \int\xi\xi^{o}d\mu, x'> = \int\xi\xi^{o}d(x'\circ\mu) =$$

$$= \int\xi d(\varphi(x'\circ\mu)\underline{\underline{R}}^{o})) = \int\varphi'\xi d(x'\circ\mu|\underline{\underline{R}}^{o}) = <\int\varphi'\xi d(\mu|\underline{\underline{R}}^{o}) \; , x'>$$

for any $x' \in E'$. Hence

$$\int \xi \xi^{\circ} d\mu = \int \varphi' \xi d(\mu | \underline{\underline{R}}^{\circ}) \ . \ \square$$

Corollary 4.2.16 Let $A \in \underline{\underline{R}}$. We set

$$\underline{\underline{R}}^{\circ} := \{B \in \underline{\underline{R}} \ , \ B \subset A\} \ ,$$

$$M^{\circ} := \{\lambda | \underline{\underline{R}}^{\circ} | \lambda \in M\}$$

and denote by $\varphi\lambda^{\circ}$ *the map*

$$\underline{\underline{R}} \longrightarrow \mathbb{R} \ , \ B \longrightarrow \lambda^{\circ}(A \cap B)$$

for any $\lambda^{\circ} \in M^{\circ}$. *Then:*

a) $\varphi\lambda^{\circ} \in M_c$ *for any* $\lambda^{\circ} \in M^{\circ}$;

b) *for any* $\xi \in M_c^{\pi}$ *the map*

$$M^{\circ} \longrightarrow \mathbb{R} \ , \ \lambda^{\circ} \longmapsto \xi(\varphi\lambda^{\circ})$$

belongs to $M^{\circ\pi}$; *we denote it by* $\varphi'\xi$;

c) *if* $\xi \in M_b^{\pi}$ *then* $\|\varphi'\xi\| \leq \|\xi\|$;

d) *we have*

$$\int \xi i_A d\mu = \int \varphi' \xi d(\mu | \underline{\underline{R}}^{\circ})$$

for any $\mu \in M(E)$ *and* $\xi \in M_c^{\pi}$.

a), b), and d) follow from Proposition 4.2.15. c) is trivial. □

Corollary 4.2.17 Let $\mu \in M(E)$, *let* $\xi \in M_c^{\pi}$, *and let* $A \in \underline{\underline{R}}$. *If* E *is* δ-*quasicomplete and* Σ-*complete then:*

a) $\int \xi i_A d\mu \in E$;

b) $\xi \cdot \mu \in M(E)$;

c) *the map*

$$M_b^{\pi} \longrightarrow E \ , \ \eta \longmapsto \int \eta i_A d\mu$$

is continuous.

a) follows from Corollary 4.2.16 b) d) and from the last assertion of Theorem 4.2.11.

b) follows immediately from a).

c) follows from Corollary 4.2.16 b), c), d) and from Theorem 4.2.11 d) and its last assertion. □

Proposition 4.2.18 *Let* $\mu \in M(E)$ *and let* $(\xi_n)_{n\in\mathbb{N}}$ *be an increasing sequence in* $\hat{\mathcal{L}}^1(\mu)$ *whose supremum* ξ *belongs to* $\hat{\mathcal{L}}^1(\mu)$ *. Then* $(\int \xi_n d\mu)_{n\in\mathbb{N}}$ *is a* Σ*-sequence converging to* $\int \xi d\mu$ *in* $E'^*_{E'}$ *.*

We have

$$<\int \xi d\mu \ , x'> = \int \xi d(x' \circ \mu) = \lim_{n\to\infty} \int \xi_n d(x' \circ \mu) = \lim_{n\to\infty} <\int \xi_n d\mu \ , x'> \ ,$$

$$\int (\xi - \xi_o) d|x' \circ \mu| = \sum_{n\in\mathbb{N}} \int (\xi_{n+1} - \xi_n) d|x' \circ \mu|$$

for any $x' \in E'$. By the first relation $(\int \xi_n d\mu)_{n\in\mathbb{N}}$ converges in $E'^*_{E'}$ to $\int f d\mu$. From the second we deduce that for any finite subset I of E' there exists a finite subset J of $\mathbb{N}$ such that

$$\left| \sum_{n\in K} <\int \xi_{n+1} d\mu - \int \xi_n d\mu \ , x'> \right| < 1$$

for any $x' \in I$ and for any finite subset K of $\mathbb{N} \setminus J$. Hence $(\int \xi_n d\mu)_{n\in\mathbb{N}}$ is a Σ-sequence in $E'^*_{E'}$. □

Theorem 4.2.19 *Assume* E δ*-quasicomplete and weakly* Σ*-complete and let* $\mu \in M(E)$ *. Then:*

a) $\int \xi d\mu \in E$ *for any* $\xi \in \hat{\mathcal{L}}^1(\mu)$ *;*

b) $\xi \cdot \mu \in M(E)$ *for any* $\xi \in \hat{\mathcal{L}}^1_{loc}(\mu)$ *;*

a) Let $\xi \in \hat{\mathcal{L}}^1(\mu) \cap M^{\pi}_c$. By Corollary 4.2.17 b) and Theorem 3.2.2 d) $\xi \cdot \mu \in M_b(E)$.

Let $\underline{\underline{S}}$ be the σ-ring generated by $\underline{\underline{R}}$. Let $A \in \underline{\underline{S}}$. There exists an increasing sequence $(A_n)_{n\in\mathbb{N}}$ in $\underline{\underline{R}}$ with $A = \bigcup_{n\in\mathbb{N}} A_n$. Then $(\dot{1}_{A_n})_{n\in\mathbb{N}}$ is an increasing sequence in $\hat{\mathcal{L}}^1(\xi\cdot\mu)$ and its supremum $\dot{1}_A$ belongs to $\hat{\mathcal{L}}^1(\xi\cdot\mu)$ too. By Proposition 4.2.18 $(\xi\cdot\mu(A_n))_{n\in\mathbb{N}}$ is a weakly Σ-sequence in E converging in $E'^{*}_{E'}$ to $\int\dot{1}_A d(\xi\cdot\mu)$. Since E is weakly Σ-complete we get $\int\dot{1}_A d(\xi\cdot\mu)\in E$.

We set

$$\nu : \underline{\underline{S}} \longrightarrow E\,,\; A \longmapsto \int\dot{1}_A d(\xi\cdot\mu)\,.$$

ν is a measure. We have

$$<\int\dot{1}d(\xi\cdot\mu)\,,\, x'> = \int 1_X d(x'\circ(\xi\cdot\mu)) = \int 1_X d(x'\circ\nu) = <\int 1_X d\nu\,,\, x'>$$

for any $x'\in E'$ and therefore by Theorem 3.2.2 d) and the last assertion of Theorem 4.2.11

$$\int\xi d\mu = \int\dot{1}d(\xi\cdot\mu) = \int\dot{1}d\nu\in E\,.$$

Let now ξ be a positive element of $\hat{\mathcal{L}}^1(\mu)$. By Corollary 3.1.2 b) we have $\xi\wedge n\dot{1}\in M_b^{\pi}$ and therefore, by the above considerations, $\int\xi\wedge n\dot{1}d\mu\in E$ for any $n\in\mathbb{N}$. Using once more the weak Σ-completeness of E we get by Proposition 4.2.18 $\int\xi d\mu\in E$.

For an arbitrary $\xi\in\hat{\mathcal{L}}^1(\mu)$ we deduce from the above result

$$\int\xi d\mu = \int\xi\vee 0 d\mu - \int(-\xi)\vee 0 d\mu\in E\,.$$

b) follows immediately from a). □

<u>*Theorem 4.2.20*</u> *Let* $\mu\in M(E)$ *and* $\xi\in\hat{\mathcal{L}}^1(\mu)$. *Then:*

a) $\{\eta\in M^{\rho} \mid |\eta| \leqslant |\xi|\}\subset\mathcal{L}^1(\mu)$;

b) the set

$$\{\int\eta d\mu \mid \eta\in M^{\rho}\,,\; |\eta| \leqslant |\xi|\}$$

is a compact set of $E'^{*}_{E'}$;

c) *if* E *is* δ*-quasicomplete and* Σ*-complete then* $\int \xi d\mu \in E''$;

d) *if* E *is* δ*-quasicomplete and weakly* σ*-complete then the set*

$$\{\int \eta d\mu \mid \eta \in M^{\rho} \;,\; |\eta| \leqslant |\xi|\}$$

is a weakly compact set of E .

a) follows from the fact that $\hat{\mathcal{L}}^1(\mu)$ is a solid subspace of M^{ρ} .

b) we denote by N the solid subspace of M generated by $\{x'\circ\mu \mid x' \in E'\}$. Then $\hat{\mathcal{L}}^1(\mu) \subset N^{\pi}$ and $\{\eta \in \hat{\mathcal{L}}^1(\mu) \mid |\xi| \leqslant |\eta|\}$ is a compact set of $\hat{\mathcal{L}}^1(\mu)$ for the weak topology associated with the duality $\langle \hat{\mathcal{L}}^1(\mu), N\rangle$ (Proposition 3.1.8 b)). The map

$$\hat{\mathcal{L}}^1(\mu) \longrightarrow E'^{*} \;,\; \xi \longmapsto \int \xi d\mu$$

is the adjoint of the map

$$E' \longrightarrow N,\; x' \longmapsto x'\circ\mu$$

and therefore it is continuous with respect to the corresponding weak topologies. Hence

$$\{\int \eta d\mu \mid \eta \in \hat{\mathcal{L}}^1(\mu) \;,\; |\eta| \leqslant |\xi|\}$$

is a compact set of $E'^{*}_{E'}$.

c) By Corollary 3.1.2 b) $(\xi \vee (-n1_A)) \wedge n1_A \in M^{\pi}_b$ and therefore by Corollary 4.2.17 a)

$$\int (\xi \vee (-n1_A)) \wedge n1_A d\mu \in E$$

for any $(n,A) \in \mathbb{N} \times \underline{\underline{R}}$. Let $\underline{\underline{F}}$ be the filter on $\mathbb{N} \times \underline{\underline{R}}$ generated by the filter base

$$\{\{(m,B) \in \mathbb{N} \times \underline{\underline{R}} \mid m \geqslant n,\; B \supset A\} \mid (n,A) \in \mathbb{N} \times \underline{\underline{R}}\} \;.$$

We have

$$\lim_{(n,A),\underline{\underline{F}}} \int (\xi \vee (n1_A)) \wedge n1_A d(x'\circ\mu) = \int \xi d(x'\circ\mu)$$

for any $x'\in E'$ (Theorem 2.3.8). Hence the image of $\underline{\underline{F}}$ with respect to the map

$$\mathbb{N}\times\underline{\underline{R}} \longrightarrow E'^{*}_{E'}\ ,\ (n,A) \longmapsto \int(\xi\vee(-n\,1_A))\wedge n\,1_A d\mu$$

converges to $\int\xi d\mu$. Hence $\int\xi d\mu$ belongs to the closure in $E'^{*}_{E'}$ of

$$E\cap\{\int\eta d\mu \mid \eta\in M^{\rho}\ ,\ |\eta|\leqslant|\xi|\}.$$

By b) this set is bounded and therefore $\int\xi d\mu\in E''$.

d) follows immediately from b) and Theorem 4.2.19 a). □

<u>Remark.</u> In general $\int\xi d\mu\notin E$ for $\mu\in M(E)$ and $\xi\in\hat{\mathcal{L}}^1(\mu)$ even if E is a Banach space. Indeed let E be c_o , X be $\mathbb{N}$, $\underline{\underline{R}}$ be the set of finite subsets of $\mathbb{N}$, and μ be

$$\underline{\underline{R}} \longrightarrow E\ ,\ A \longmapsto 1_A\ .$$

Then $1\in\hat{\mathcal{L}}^1(\mu)$ and $\int 1 d\mu = 1_X\in\ell^{\infty} = c_o''$ but $1_X\notin c_o$.

<u>*Definition 4.2.21*</u> *For any* $\lambda\in M$ *we denote by* $\hat{\mathcal{L}}^1_{loc}(\lambda,E)$ *the set of linear maps* φ *of* E' *into* $\hat{\mathcal{L}}^1_{loc}(\lambda)$ *such that for any* $A\in\underline{\underline{R}}$ *the map*

$$E' \longrightarrow \mathbb{R}\ ,\ x' \Longrightarrow ((\varphi x')\cdot\lambda)(A)$$

is continuous with respect to the $\sigma(E',E)$*-topology. For any* $A\in\underline{\underline{R}}$, $\lambda\in M$, *and* $\varphi\in\hat{\mathcal{L}}^1_{loc}(\lambda,E)$ *we denote by* $(\varphi\cdot\lambda)(A)$ *the element of* E *such that*

$$((\varphi x')\cdot\lambda)(A) = <(\varphi\cdot\lambda)(A)\ ,\ x'>$$

for any $x'\in E'$ *. With this notation we have*

$$x'\circ(\varphi\cdot\lambda) = \varphi(x')\cdot\lambda$$

for any $x'\in E'$ *and therefore* $\varphi\cdot\lambda\in M(E)$ *for any* $\lambda\in M$ *and* $\varphi\in\hat{\mathcal{L}}^1_{loc}(\lambda,E)$. *We set*

$$\hat{L}^1_{loc}(\lambda,E) := \{\varphi\in\hat{\mathcal{L}}^1_{loc}(\lambda) \mid \varphi(E')\subset\hat{L}^1_{loc}(\mu)\}$$

for any $\lambda \in M$. *For any* $\lambda \in M$ *and any* $(\xi,\varphi) \in M^{\rho} \times L^{1}_{loc}(\lambda)$ *we denote by* $\xi\varphi$ *the map*

$$E' \longrightarrow M^{\rho} , \quad x' \longmapsto \xi\varphi(x')$$

Proposition 4.2.22 *Let* $\lambda \in M$ *and let* M_{λ} *be the set of elements of* $M(E)$ *which are absolutely continuous with respect to* λ . *Then :*

a) M_{λ} *is a subspace of* $M(E)$;

b) $\varphi \cdot \lambda \in M_{\lambda}$ *for any* $\varphi \in \hat{L}^{1}_{loc}(\lambda,E)$;

c) *the map*

$$\hat{L}^{1}_{loc}(\lambda,E) \longrightarrow M_{\lambda} , \quad \varphi \longrightarrow \varphi \cdot \lambda$$

is linear and surjective ;

d) *the map*

$$\hat{L}^{1}_{loc}(\lambda,E) \longrightarrow M_{\lambda} , \quad \varphi \longrightarrow \varphi \cdot \lambda$$

is an isomorphism of vector spaces ;

e) *if* E *is* δ*-quasicomplete and weakly* Σ*-complete then*

$$\xi \in \hat{L}^{1}_{loc}(\lambda,E) \Longleftrightarrow \xi u \in \hat{L}^{1}_{loc}(\lambda) \Longrightarrow \xi \cdot (\varphi \cdot \lambda) = (\xi\varphi) \cdot \lambda$$

for any $(\xi,\varphi) \in M^{\rho} \times \hat{L}^{1}_{loc}(\lambda)$.

a) and b) are obvious.

c & d. It is clear that the maps are linear. Let $\nu \in M_{\lambda}$. By Theorem 3.2.2 c) there exists a unique map

$$\varphi : E' \longrightarrow \hat{L}^{1}_{loc}(\lambda)$$

such that $\varphi(x') \cdot \lambda = x' \circ \nu$ for any $x' \in E'$. It is obvious that u is linear. We have

$$\langle (\varphi\lambda)(A), x' \rangle = ((\varphi x') \cdot \lambda)(A) = (x' \circ \nu)(A) = \langle \nu(A) , x' \rangle$$

for any $(x',A) \in E' \times \underline{\underline{R}}$ and therefore $\varphi \in \hat{L}^{1}_{loc}(\lambda,E)$ and $\varphi \cdot \lambda = \nu$.

This proves that both maps are surjective. By Theorem 3.2.2 c) the map from d) is bijective.

e) follows immediately from Theorem 3.2.2 d) and Theorem 4.2.19 b). □

The following lemma is proved for later use.

Lemma 4.2.23 *We set*

$$\ell^{\infty}(E) := \{(x_n)_{n\in\mathbb{N}} \in E^{\mathbb{N}} \mid \{x_n\}_{n\in\mathbb{N}} \text{ bounded set of } E\},$$

$$c_o(E) := \{(x_n)_{n\in\mathbb{N}} \in E^{\mathbb{N}} \mid \lim_{n\to\infty} x_n = 0\}$$

and denote by P *the set of continuous seminorms* p *on* E *and for any* $p\in P$ *by* p_o *the map*

$$\ell^{\infty}(E) \longrightarrow \mathbb{R}_+ , \quad (x_n)_{n\in\mathbb{N}} \longmapsto \sup_{n\in\mathbb{N}} p(x_n) .$$

We denote by $\underline{\underline{U}}$ *the set of circled convex neighbourhoods in* E *and for any* $U\in\underline{\underline{U}}$ *by* U^o *its polar set in* E' *and set*

$$\ell^{1}(E) := \{(x'_n)_{n\in\mathbb{N}} \in E'^{\mathbb{N}} \mid \exists U\in\underline{\underline{U}},\ \exists(\alpha_n)_{n\in\mathbb{N}} \in \ell^{1}, (\forall n\in\mathbb{N} \Longrightarrow x'_n \in \alpha_n U^o)\}$$

Then:

a) $\ell^{\infty}(E)$ *is a subspace of* $E^{\mathbb{N}}$ *, and* p_o *is a seminorm for any* $p\in P$ *; we endow* $\ell^{\infty}(E)$ *with the topology generated by the family* $(p_o)_{p\in P}$ *of seminorms ;* $\ell^{\infty}(E)$ *is complete if* E *is complete ;*

b) $c_o(E)$ *is a closed subspace of* $\ell^{\infty}(E)$ *;*

c) the sequence $(\langle x_n, x'_n\rangle)_{n\in\mathbb{N}}$ *is summable for any* $(x_n)_{n\in\mathbb{N}} \in \ell^{\infty}(E)$ *and* $(x'_n)_{n\in\mathbb{N}} \in \ell^{1}(E)$ *and the maps*

$$\ell^{\infty}(E)\times\ell^{1}(E) \longrightarrow \mathbb{R} , \quad ((x_n)_{n\in\mathbb{N}}, (x'_n)_{n\in\mathbb{N}}) \longmapsto \sum_{n\in\mathbb{N}} \langle x_n, x'_n\rangle ,$$

$$c_o(E)\times\ell^{1}(E) \longrightarrow \mathbb{R} , \quad ((x_n)_{n\in\mathbb{N}}, (x'_n)_{n\in\mathbb{N}}) \longmapsto \sum_{n\in\mathbb{N}} \langle x_n, x'_n\rangle$$

are separated dualities ;

d) for any $U \in \underline{\underline{U}}$ *the set*

$$S := \{(x'_n)_{n\in\mathbb{N}} \in \ell^1(E) \mid \exists (\alpha_n)_{n\in\mathbb{N}} \in \ell^1, \sum_{n\in\mathbb{N}} |\alpha_n| \leq 1, (\forall n\in\mathbb{N} \Longrightarrow x'_n \in \alpha_n U^\circ)\}$$

is compact for the $\sigma(\ell^1(E), c_o(E))$*-topology ;*

e) the topology induced on $c_o(E)$ *by* $\ell^\infty(E)$ *is coarser than the* $\tau(c_o(E), \ell^1(E))$*-topology ;*

f) if $(\mu_n)_{n\in\mathbb{N}}$ *is a sequence in* $M(E)$ *such that*

$$(\mu_n(A))_{n\in\mathbb{N}} \in \ell^\infty(E)$$

for any $A \in \underline{\underline{R}}$ *and if we denote by* F *the space* $\ell^\infty(E)$ *endowed with the* $\sigma(\ell^\infty(E), \ell^1(E))$*-topology then*

$$\underline{\underline{R}} \longrightarrow \ell^\infty(E), \; A \longmapsto (\mu_n(A))_{n\in\mathbb{N}}$$

belongs to $M(F)$ *;*

g) if $(\mu_n)_{n\in\mathbb{N}}$ *is a sequence in* $M(E)$ *such that*

$$(\mu_n(A))_{n\in\mathbb{N}} \in c_o(E)$$

for any $A \in \underline{\underline{R}}$ *and if we denote by* F *the space* $c_o(E)$ *endowed with the topology induced by* $\ell^\infty(E)$ *then*

$$\underline{\underline{R}} \longrightarrow c_o(E), \; A \longmapsto (\mu_n(A))_{n\in\mathbb{N}}$$

belongs to $M(F)$ *.*

a), b), and c) are trivial.

d) We denote for any $n\in\mathbb{N}$ by φ_n the map

$$S \longrightarrow U^\circ, \; (x'_n)_{n\in\mathbb{N}} \longmapsto x'_n .$$

Let $\underline{\underline{F}}$ be an ultrafilter on S . Then $\varphi_n(\underline{\underline{F}})$ converges for any $n\in\mathbb{N}$ with respect to the topology induced on U° by $\sigma(E',E)$; we denote by y'_n its limit. Let $(x_n)_{n\in\mathbb{N}} \in c_o(E)$. Then

$$\sum_{n\in M} \langle x_n , y'_n \rangle = \lim_{(x'_n)_{n\in\mathbb{N}}, \underline{\underline{F}}} \sum_{n\in M} \langle x_n , x'_n \rangle$$

and therefore

$$\left| \sum_{n\in M} \langle x_n , y'_n \rangle \right| \leqslant \sup_{n\in M} p(x_n)$$

for any finite subset M of $\mathbb{N}$, where p denotes the seminorm on E associated with U. We deduce $(\langle x_n , y'_n \rangle)_{n\in\mathbb{N}}$ is summable and

$$\sum_{n\in\mathbb{N}} \langle x_n , y'_n \rangle = \lim_{(x'_n)_{n\in\mathbb{N}}, \underline{\underline{F}}} \sum_{n\in\mathbb{N}} \langle x_n , x'_n \rangle .$$

Since $(x_n)_{n\in\mathbb{N}}$ is arbitrary $(y'_n)_{n\in\mathbb{N}} \in S$ and $\underline{\underline{F}}$ converges to $(y'_n)_{n\in\mathbb{N}}$ with respect to $\sigma(\ell^1(E) , c_o(E))$.

e) Let $U \in \underline{\underline{U}}$, let p be the corresponding seminorm on E, and let S be the set defined in d). Then the polar of S in $c_o(E)$ is the set $\{x \in c_o(E) \mid p_o(x) \leqslant 1\}$. By d) the topology induced by $\ell^\infty(E)$ on $c_o(E)$ is coarser than $\tau(c_o(E) , \ell^1(E))$.

f) By Nikodym's theorem on bounded measures ([8] Theorem I 3.1)

$$\{\mu_n(A) \mid n\in\mathbb{N} , A\in\underline{\underline{R}} , A\subset B\}$$

is a bounded set of E for any $B\in\underline{\underline{R}}$. The assertion follows immediately from this remark.

g) The assertion follows immediately from e) and f). □

3. Operators on subspaces of M_c^π

Proposition 4.3.1 *Let* F *be a solid subspace of* M_c^π *containing* $\{\dot{1}_A \mid A\in\underline{\underline{R}}\}$, *let* N *be the set* $\bigcap_{\xi\in F} M(\xi)$, *and let* φ *be a linear map* $F \longrightarrow E$. *Then* N *is a fundamental solid subspace of* M, $F\subset N^\pi$, *and the following assertions are equivalent:*

a) φ *is continuous with respect to the Mackey topology* $\tau(F,N)$;

b) φ *is order continuous* ;

c) *there exists* $\mu \in N(E)$ *such that*

$$\varphi(\xi) = \int \xi d\mu$$

for any $\xi \in F$.

Each of the above assertions implies

$$\varphi(\{\eta \in F \mid |\eta| \leqslant |\xi|\})$$

is a weakly compact set of E *for any* $\xi \in F$.

It is obvious that N is a fundamental solid subspace of M and $F \subset N^{\pi}$.

a $\Longrightarrow$ b . Let A be a nonempty lower directed set in F with infinum 0 and let $\underline{\underline{F}}$ be the filter on F generated by the filter base

$$\{\{y \in A \mid y \leqslant x\} \mid x \in A\} .$$

Then $\underline{\underline{F}}$ converges to 0 with respect to the $\sigma(F,N)$ topology and therefore $\varphi(\underline{\underline{F}})$ weakly converges to 0 in E . Hence φ is order continuous with respect to the weak topology of E . By Proposition 4.1.6 φ is order continuous.

b $\Longrightarrow$ c . We denote by μ the map

$$\underline{\underline{R}} \longrightarrow E , A \longmapsto \varphi(i_A) .$$

For any $x' \in E'$ we have $x' \circ \varphi \in F^{\pi}$. By Proposition 3.3.9 there exists for any $x' \in E'$ a $\lambda_{x'} \in N$ such that

$$\xi(\lambda_{x'}) = x' \circ \varphi(\xi)$$

for any $\xi \in F$. We get $\lambda_{x'} = x' \circ \mu$ for any $x' \in E'$ and therefore $\mu \in N(E)$. We have

$$\langle \int \xi d\mu , x' \rangle = \int \xi d(x' \circ \mu) = \int \xi d\lambda_{x'} = x' \circ \varphi(\xi) = \langle \varphi(\xi) , x' \rangle$$

for any $\xi \in F$ and for any $x' \in E'$. Hence

$$\varphi(\xi) = \int \xi d\mu$$

for any $\xi \in F$.

c $\Rightarrow$ a . The map φ is the adjoint of the map

$$E' \longrightarrow N , \quad x' \longmapsto x' \circ \mu$$

and therefore it is continuous with respect to the Mackey topology $\tau(F,N)$.

We prove now the last assertion. Let $\xi \in F$. By Proposition 3.1.8 $\{\eta \in F \mid |\eta| \leq |\xi|\}$ is compact with respect to the $\sigma(F,N)$ topology. By a) φ is continuous with respect to the $\sigma(F,N)$ and $\sigma(E,E')$ topologies Hence $\varphi(\{\eta \in F \mid |\eta| \leq |\xi|\})$ is weakly compact. □

Corollary 4.3.2 *If* φ *is an order continuous linear map* $M_b^\pi \longrightarrow E$ *and if* ψ *is an order continuous linear map* $M_b^\pi \longrightarrow M_b^\pi$ *then* $\{\varphi \circ \psi(\xi) \mid \xi \in M_b^\pi , \|\xi\| \leq 1\}$ *is a compact set of* E .

The assertion follows immediately from Proposition 4.3.1 and Proposition 3.8.5 c $\Rightarrow$ b . □

Corollary 4.3.4 *Let* Y *be a locally compact hyperstonian space, let* F *be a solid subspace of* $C(Y)$ *containing* $C_c(Y)$ *, and let* φ *be a linear map* $F \longrightarrow E$ *. Then the following assertions are equivalent:*

a) φ *is order continuous ;*

b) there exists $\mu \in (M(Y))(E)$ *such that* $F \subset L^1(x' \circ \mu)$ *for any* $x' \in E'$ *and*

$$\varphi(f) = \int f d\mu$$

for any $f \in F$.

By Proposition 2.3.9 there exists an isomorphism of vector lattices $u\colon C(Y) \longrightarrow M(Y)_c^\pi$ such that

$$f \in L^1(\lambda) \Longleftrightarrow uf \in \hat{L}^1(\lambda) \Longrightarrow \int f d\lambda = \int u(f) d\lambda$$

for any $\lambda \in M(Y)$. We set $N := \bigcap_{\xi \in u(F)} M(\xi)$.

a $\Longrightarrow$ b follows immediately from Proposition 4.3.1 b $\Longrightarrow$ c .

b $\Longrightarrow$ a . Since $F \subset \mathcal{L}^1(x' \circ \mu)$ for any $x' \in E'$ we have $\mu \in N(E)$. Moreover

$$\varphi(u^{-1}(\xi)) = \int \xi d\mu$$

for any $\xi \in u(F)$. By Proposition 4.3.1 c $\Longrightarrow$ b $\varphi \circ u^{-1}$ is order continuous. Hence φ is order continuous.□

Theorem 4.3.5 *Let* F *be the set* $\{\dot{1}_A | A \in \underline{\underline{R}}\}$ *and let* $\mu \in M_c(E)$ *such that* $\int \xi d\mu \in E$ *for any* $\xi \in M_c^{\pi}$. *Then* $\{\int \xi d\mu | \xi \in A\}$ *is precompact for any weakly pseudo-compact set* A *of* (M_c^{π}, F) *and it is compact for any weakly compact set* A *of* (M_c^{π}, F) .

Let us denote by φ the map

$$M_c^{\pi} \longrightarrow E \ , \ \xi \longmapsto \int \xi d\mu \ .$$

This map is continuous with respect to the Mackey topology $\tau(M_c^{\pi}, M_c)$ (Proposition 4.3.1 c $\Longrightarrow$ a). The assertions follow from Corollary 3.8.6.□

Proposition 4.3.6 *Let* $\mu \in M(E)$, *let* $\xi \in \hat{\mathcal{L}}^1(\mu)$, *let* U *be a circled convex 0-neighbourhood, and let* U^o *be its polar set in* E'. *Then :*

a) $\sup_{x' \in U^o} \int |\xi| d|x' \circ \mu| < \infty$;

b) if $\int \eta d\mu \in E$ *for any* $\eta \in \hat{\mathcal{L}}^1(\mu)$ *with* $|\eta| \leq |\xi|$ *then*

$$\sup_{x' \in U^o} \int |\xi| d|x' \circ \mu| = \inf\{\alpha \in \mathbb{R}_+ | A \subset \alpha U\} ,$$

where $A := \{\int \eta d\mu | \eta \in \hat{\mathcal{L}}^1(\mu), |\eta| \leq |\xi|\}$.

a) We may assume E complete. By Theorem 4.2.20 c) $\int \eta d\mu \in E''$ for any $\eta \in \hat{\mathcal{L}}^1(\mu)$. By Theorem 4.2.20 b) the set

$$\{\int \eta d\mu \mid \eta \in \hat{\mathcal{L}}^1(\mu) \;,\; |\eta| \leq |\xi|\}$$

is a compact set of $E''_{E'}$ and therefore it is equibounded on U°. The assertion follows from the relation

$$\int |\xi| d|x' \circ \mu| = \sup \{\int \eta \, d(x' \circ \mu) \mid \eta \in \hat{\mathcal{L}}^1(\mu) \;,\; |\eta| \leq |\xi|\},$$

which holds for any $x' \in E'$.

b) Let $\eta \in \hat{\mathcal{L}}^1(\mu)$ with $|\eta| \leq |\xi|$ and let $x' \in U^\circ$. Then

$$|\langle \int \eta d\mu, x' \rangle| = |\int \eta d(x' \circ \mu)| \leq \int |\xi| d|x' \circ \mu|$$

and therefore

$$\inf \{\alpha \in \mathbb{R}_+ \mid A \subset \alpha U\} \leq \sup_{x' \in U^\circ} \int |\xi| d|x' \circ \mu| .$$

Let now $\alpha \in \mathbb{R}_+$ such that $A \subset \alpha U$ and let $x' \in U^\circ$. We have

$$\int |\xi| d|x' \circ \mu| = \sup \{\int \eta d(x' \circ \mu) \mid \eta \in \hat{\mathcal{L}}^1(\mu) \;,\; |\eta| \leq |\xi|\}$$

$$= \sup_{x \in A} \langle x, x' \rangle \leq \alpha$$

and therefore

$$\sup_{x' \in U^\circ} \int |\xi| d|x' \circ \mu| \leq \inf \{\alpha \in \mathbb{R}_+ \mid A \subset \alpha U\} . \quad \square$$

Corollary 4.3.7 *Let* $\mu \in M(E)$ *and let* $\xi \in \hat{\mathcal{L}}^1(\mu)$. *We set*

$$F := \{\eta \in \hat{\mathcal{L}}^1(\mu) \mid \exists n \in \mathbb{N} \;,\; |\eta| \leq n|\xi| \;,\; \int \eta d\mu \in E\}$$

and endow F *with the norm*

$$F \longrightarrow \mathbb{R} \;,\; \eta \longmapsto \inf \{\alpha \in \mathbb{R}_+ \mid \; |\eta| \leq \alpha |\xi|\}.$$

Then the map

$$F \longrightarrow E\,,\ \eta \longmapsto \int \eta d\mu$$

is continuous.

Let U be a closed convex 0-neighbourhood in E and let U° be its polar set in E'. By Proposition 4.3.6 a)

$$\alpha := \sup_{x'\in U^\circ} \int |\xi| d|x' \circ \mu| < \infty .$$

We get

$$|\langle \int \eta d\mu , x'\rangle| \leq \int |\eta| d|x' \circ \mu| \leq \alpha \|\eta\|$$

for any $\eta \in F$ and for any $x' \in U^\circ$. Hence

$$\int \eta d\mu \in \|\eta\| U$$

for any $\eta \in F$ and the map

$$F \longrightarrow E\,,\ \eta \longmapsto \int \eta d\mu$$

is continuous. □

Proposition 4.3.8 *Let $\mu \in M(E)$, let $\xi \in M_c^\pi$, and let F be the solid subspace of M_c^π generated by $\{i_A | A \in \underline{\underline{R}}\}$. If E is δ-quasicomplete and Σ-complete and if there exists $\eta \in \mathcal{L}^1(\mu)$ such that $(n|\xi|-|\eta|) \vee 0 \in F$ for any $n \in \mathbb{N}$ then $\xi \in \mathcal{L}^1(\mu)$ and $\int \xi d\mu \in E$.*

We may construct inductively an increasing sequence $(A_n)_{n\in\mathbb{N}}$ in $\underline{\underline{R}}$ such that for any $n \in \mathbb{N}$, $(2^n|\xi|-|\eta|) \vee 0$ belongs to the solid subspace of M_c^π generated by i_{A_n}. By Corollary 4.2.17 a) $\int \xi i_{A_n} d\mu \in E$ for any $n \in \mathbb{N}$. We want to show that $(\int \xi i_{A_n} d\mu)_{n\in\mathbb{N}}$ is a Σ-sequence in E. We have successively

$$(2^n|\xi|-|\eta|) \vee 0 = ((2^n|\xi|-|\eta|) \vee 0) i_{A_n} ,$$

$$(2^n|\xi|) \vee |\eta| - |\eta| = ((2^n|\xi|) \vee |\eta| - |\eta|) i_{A_n} ,$$

$$2^n|\xi|(1-i_{A_n}) \leq ((2^n|\xi|)\vee|\eta|)(1-i_{A_n}) \leq |\eta|(1-i_{A_n}) \leq |\eta| ,$$

$$|\xi| i_{X\setminus A_n} \leq \frac{1}{2^n}|\eta|$$

for any $n\in\mathbb{N}$. Let U be a closed convex 0-neighbourhood in E and let U^0 be its polar set in E' . By Proposition 4.3.6 a)

$$\alpha := \sup_{x'\in U^0} \int|\eta| d|x'\circ\mu| < \infty .$$

We get

$$\sup_{x'\in U^0} \int|\xi|(i_{A_n} - 1_{A_m}) d|x'\circ\mu| \leq \frac{\alpha}{2^m}$$

and therefore

$$\int\xi i_{A_n} d\mu - \int\xi 1_{A_m} d\mu \in \frac{\alpha}{2^m} U$$

for any $m, n\in\mathbb{N}$ with $m \leq n$. This shows that $\xi\in\hat{\mathcal{L}}^1(\mu)$ and that $(\int\xi i_{A_n} d\mu)_{n\in\mathbb{N}}$ is a Σ-sequence and therefore a convergent sequence in E .

Let $x'\in E'$. We have

$$|\int\xi d(x'\circ\mu) - \int\xi i_{A_n} d(x'\circ\mu)| \leq \int|\xi| i_{X\setminus A_n} d|x'\circ\mu| \leq$$

$$\leq \frac{1}{2^n} \int|\eta| d|x'\circ\mu|$$

for any $n\in\mathbb{N}$ and therefore

$$\int\xi d(x'\circ\mu) = \lim_{n\to\infty} \int\xi i_{A_n} d(x'\circ\mu) .$$

Hence $(\int\xi i_{A_n} d\mu)_{n\in\mathbb{N}}$ converges in $E'^{*}_{E'}$ to $\int\xi d\mu$ and we deduce $\int\xi d\mu\in E$. □

Proposition 4.3.9 Let F be a vector sublattice of M_C^{π} such that $\dot{i}_A$ belongs to the solid subspace G of M_C^{π} generated by F for any $A \in \underline{\underline{R}}$, let N be the solid subspace $\bigcap_{\xi \in F} M(\xi)$ of M, and let φ be a linear map $F \longrightarrow E$. Let us consider the following assertions:

1) φ is continuous with respect to the Mackey topology $\tau(F,N)$ and

$$\{\varphi(\eta) \mid \eta \in F, |\eta| \leqslant |\xi|\}$$

is weakly relatively compact for any $\xi \in F$;

2) there exists $\mu \in N(E)$ such that

$$\varphi(\xi) = \int \xi d\mu$$

for any $\xi \in F$ and $\int \xi d\mu \in E$ for any $\xi \in G$;

3) there exists $\mu \in N(E)$ such that

$$\varphi(\xi) = \int \xi d\mu$$

for any $\xi \in F$.

If

$$(\lambda|F)^{+} = \lambda^{+}|F$$

for any $\lambda \in N$ then

a) $1 \Longleftrightarrow 2 \Longrightarrow 3$;

b) if E is δ-complete and weakly Σ-complete then $3 \Longrightarrow 2$;

c) if E is δ-complete and Σ-complete and if for any $\xi \in F$ there exists $\eta \in \hat{L}^{1}(\mu)$ such that $(n|\xi| - |\eta|) \vee 0$ belongs to the solid subspace of M_C^{π} generated by $\{\dot{i}_A | A \in \underline{\underline{R}}\}$ then $3 \Longrightarrow 2$.

a) $1 \Longrightarrow 2$. Let φ' be the adjoint map

$$E' \longrightarrow N, \quad x' \longmapsto x' \circ \varphi$$

of φ . Let $\xi \in F$, let A be the set $\{\eta \in F \mid |\eta| \leqslant |\xi|\}$, and let A° be the polar set of A in N . For any $\lambda \in N$ we have

$$q_{\xi}(\lambda) = \int |\xi| d|\lambda| = \sup \{\lambda(\eta) \mid \eta \in M_{C}^{\pi}, |\eta| \leqslant |\xi|\} =$$

$$= \sup_{\eta \in A} \lambda(\eta)$$

Hence

$$A^{\circ} = \{\lambda \in N \mid q_{\xi}(\lambda) \leqslant 1\} ,$$

and therefore

$$\overset{-1}{\varphi'}(\{\lambda \in N \mid q_{\lambda}(\xi) \leqslant 1\}) = \overset{-1}{\varphi'}(A^{\circ}) = \varphi(A)^{\circ} ,$$

where $\varphi(A)^{\circ}$ denotes the polar set of $\varphi(A)$ in E'. Hence φ' is continuous with respect to the topologies (N,F) and $\tau(E',E)$. Let G be the solid subspace of M_{C}^{π} generated by F . By Proposition 3.4.2 G is the dual of (N,F) . Hence the adjoint of φ' is a map $\varphi'': G \longrightarrow E$ which is continuous with respect to the $\tau(G,N)$-topology. By Proposition 4.3.1 a $\Longrightarrow$ c there exists $\mu \in N(E)$ such that

$$\varphi''(\xi) = \int \xi d\mu$$

for any $\xi \in G$. We get

$$\varphi(\xi) = \int \xi d\mu$$

for any $\xi \in F$.

2 $\Longrightarrow$ 1 follows from Theorem 4.2.20 b) and Proposition 4.3.1 c $\Longrightarrow$ a.

2 $\Longrightarrow$ 3 is trivial.

b) follows from Theorem 4.2.20 d).

c) follows from Proposition 4.3.8. □

Proposition 4.3.10 *Let* $\mu \in M(E)$ *such that there exists an* $\{x' \circ \mu \mid x' \in E'\}$*-concassage* $\underline{\underline{C}}$. *We denote by* F *the solid subspace of*

M^{π} generated by $\{i_C | C\in \underline{\underline{C}}\}$ endowed with the order topology and assume $\int \xi d\mu \in E$ for any $\xi\in F$. Then:

a) $\{\int \xi d\mu | \xi\in A\}$ is a precompact set of E for any weakly pseudo-compact set A of F;

b) $\{\int \xi d\mu | \xi\in A\}$ is a compact set of E for any weakly compact set A of F.

We set

$$\underline{\underline{R}}^{o} := \{A\in \underline{\underline{R}} \,|\, \exists \underline{\underline{C}}_o \text{ finite subset of } \underline{\underline{C}}, \; A\subset \bigcup_{C\in \underline{\underline{C}}_o} C\},$$

$$M^{o} := \{\lambda | \underline{\underline{R}}^{o} \,|\, \lambda\in M\}$$

and denote for any $\lambda^{o}\in M^{o}$ by $\varphi\lambda^{o}$ the map

$$\underline{\underline{R}} \longrightarrow \mathbb{R}, \; A \longmapsto \int 1_A d\lambda^{o}$$

and for any $\xi\in F$ by $\varphi'\xi$ the map

$$M^{o} \longrightarrow \mathbb{R}, \; \lambda^{o} \longmapsto \xi(\varphi\lambda^{o})$$

(Proposition 4.2.15 b), c)). By Proposition 4.2.15 d)

$$\int \xi d\mu = \int \varphi'\xi d(\mu | \underline{\underline{R}}^{o})$$

for any $\xi\in F$.

By Proposition 3.8.8 b), c) $\{\varphi'\xi | \xi\in F\} = M^{o\pi}$ and therefore by Proposition 3.8.8 d) F possesses the strong D.-P.-property and for any bounded set A of F there exists $\xi\in F$ with

$$A\subset \{\eta\in F | \; |\eta| \leqslant \xi\}.$$

By Theorem 4.2.20 we deduce further that $\{\int \xi d\mu | \xi\in A\}$ is a weakly relatively compact set of E for any bounded set A of F. In particular the map

$$F \longrightarrow E, \; \xi \longmapsto \int \xi d\mu$$

is continuous. The assertions follow from Lemma 3.8.3 c $\Longrightarrow$ a & b. □

Corollary 4.3.11 *We assume there exists a subset* $\underline{\underline{C}}$ *of* $\underline{\underline{R}}$ *such that the sets of* $\underline{\underline{C}}$ *are pairwise disjoint and such that for any* $A \in \underline{\underline{R}}$ *there exists a finite subset* $\underline{\underline{C}}_o$ *of* $\underline{\underline{C}}$ *with* $A \subset \bigcup_{C \in \underline{\underline{C}}_o} C$. *Let* $\mu \in M(E)$ *such that* $\int \xi d\mu \in E$ *for any* $\xi \in M^{\pi}$ *and let* M^{π} *be endowed with the order topology. Then:*

a) $\{\int \xi d\mu \mid \xi \in A\}$ *is a precompact set of* E *for any weakly pseudo-compact set* A *of* M^{π} ;

b) $\{\int \xi d\mu \mid \xi \in A\}$ *is a compact set of* E *for any weakly compact set* A *of* M^{π} .

Let $\xi \in M^{\pi}$. Assume $\xi \neq \xi i_A$ for any $A \in \underline{\underline{R}}$. Then there exists an infinite subset $\underline{\underline{C}}_o$ of $\underline{\underline{C}}$ such that $\xi i_C \neq 0$ for any $C \in \underline{\underline{C}}_o$. For any $C \in \underline{\underline{C}}_o$ there exists $\lambda_C \in M_+$ with $\int |\xi| i_C d\lambda_C = 1$. We set

$$\lambda : \underline{\underline{R}} \longrightarrow \mathbb{R} , \; A \longmapsto \sum_{C \in \underline{\underline{C}}_o} \lambda_C(A) .$$

Then $\lambda \in M_+$ and

$$\int |\xi| d\lambda = \sum_{C \in \underline{\underline{C}}_o} \int |\xi| d\lambda_C = \infty$$

and this is a contradiction. Hence there exists $A \in \underline{\underline{R}}$ with $\xi = \xi i_A$.

Since $\underline{\underline{C}}$ is an $\{x' \circ \mu \mid x' \in E'\}$-concassage the assertions follow from Proposition 4.3.10. □

4. Vector measures on Hausdorff spaces

Throughout this section we denote by X *a Hausdorff topological space, by* $\underline{\underline{R}}$ *the* δ*-ring of relatively compact Borel sets of* X, *by* $\underline{\underline{L}}$ *the set of compact sets of* X *which either are* G_δ*-sets of* X *or are not contained in compact* G_δ*-sets of* X, *and by* M *the set of real measures* λ *on* $\underline{\underline{R}}$ *such that*

$$|\lambda|(A) = \sup_{\substack{K\subset A \\ K\text{-compact}}} |\lambda|(K)$$

for any $A\in\underline{\underline{R}}$.

M is a band of the vector lattice of all real measures on $\underline{\underline{R}}$.

Theorem 4.4.1 *Let* F *be the solid subspace of* M^π *generated by* $\{i_A \mid A\in\underline{\underline{R}}\}$, *let* $\mu\in M(E)$, *and let* F *be a subspace of* E. *We assume:*

1) F *is* δ*-quasicomplete and* Σ*-complete with respect to the Mackey topology* $\tau(F',F)$ *of* F;

2) $\mu(L)\in F$ *for any* $L\in\underline{\underline{L}}$.

Then $\int\xi d\mu\in F$ *for any* $\xi\in F$ *(in particular* $\mu(\underline{\underline{R}})\subset F$*)*.

a) Assume first X compact. Then $\underline{\underline{R}}$ is a σ-ring and by Proposition 4.2.7 d) μ is bounded. We set

$$G := \{f\in L_b \mid \int f d\mu \in F\}$$

and denote by ψ the map

$$G \longrightarrow F\,,\ f \longmapsto \int f d\mu$$

and by C the set of continuous real functions on X.

1st *step* : $C\subset G$.

Let $f\in C$. We may assume $0 \leqslant f \leqslant 1$. We set

$$f_n := \frac{1}{2^n} \sum_{m=1}^{2^n} 1\{f \geq \frac{m}{2^n}\}$$

for any $n \in \mathbb{N}$. Then $(f_n)_{n\in\mathbb{N}}$ is an increasing sequence in G such that

$$0 \leq f - f_n \leq \frac{1}{2^n}$$

for any $n \in \mathbb{N}$. $(f_n)_{n\in\mathbb{N}}$ is a Σ-sequence in L_b with respect to the supremum norm on L_b and therefore by Proposition 4.2.5 d) $(\int f_n d\mu)_{n\in\mathbb{N}}$ is a Σ-sequence in F . Hence $(\int f_n d\mu)_{n\in\mathbb{N}}$ converges in F to an x . We have

$$\langle x,x'\rangle = \lim_{n\to\infty} \langle \int f_n d\mu \ , x'\rangle = \lim_{n\to\infty} \int f_n d(x' \circ \mu) =$$

$$= \int f d(x' \circ \mu) = \langle \int f d\mu \ , x'\rangle$$

for any $x' \in E'$ and therefore

$$\int f d\mu = x \in F \ .$$

2nd *step : if* $(f_n)_{n\in\mathbb{N}}$ *is an increasing upper bounded sequence in* G *then* $(\psi(f_n))_{n\in\mathbb{N}}$ *is a* Σ*-sequence in* F .

Let U be a closed convex circled 0-neighbourhood in F and let U° be its polar set in F' endowed with the induced $\sigma(F',F)$ topology. We denote by $C(U^\circ)$ the vector space of continuous real functions on U° endowed with the supremum norm. For any $f \in C$ we denote by $\varphi(f)$ the map

$$U^\circ \longrightarrow \mathbb{R} \ , \ x' \longmapsto \langle \int f d\mu \ , x'\rangle \ .$$

φ is a linear map of C into $C(U^\circ)$ and by Proposition 4.2.5 d) it is continuous.

By [14] Théorème 6, (2) $\Longrightarrow$ (1) , φ is weakly compact and therefore ([14] Lemma 1, (1) $\Longrightarrow$ (2)) $\varphi''(M_b^\pi) \subset C(U^\circ)$, where φ'' denotes the biadjoint of φ . With other words the map

$$\theta\xi : U^{o} \longrightarrow \mathbb{R} , x' \longmapsto \int \xi d(x' \circ \mu)$$

is continuous for any $\xi \in M_b^{\pi}$.

$(f_{n+1}-f_n)_{n\in\mathbb{N}}$ is a sequence of positive functions in G such that

$$X \longrightarrow \mathbb{R} , x \longmapsto \sum_{n\in\mathbb{N}} (f_{n+1}(x)-f_n(x))$$

is bounded. By the above considerations

$$(\sum_{\substack{n\in M\\ n\leq m}} \theta(\dot{f}_{n+1}-\dot{f}_n))_{m\in\mathbb{N}}$$

is weakly convergent in $C(U^{o})$ for any $M \subset N$. By Lemma 4.1.4 $(\theta(\dot{f}_{n+1}-\dot{f}_n))_{n\in M}$ is weakly summable for any $M \subset N$. By Orlicz-Pettis theorem $(\theta(\dot{f}_{n+1}-\dot{f}_n))_{n\in\mathbb{N}}$ is summable in $C(U^{o})$ and therefore $(\theta\dot{f}_n)_{n\in\mathbb{N}}$ is a Σ-sequence in $C(U^{o})$. Since U is arbitrary $(\psi(f_n))_{n\in\mathbb{N}}$ is a Σ-sequence in F .

3rd *step* : $\mu(K)\in F$ *for any compact set* K *of* X .

Let $\underline{\underline{L}}_o$ be the set of compact G_δ-sets of X containing K and let $\underline{\underline{F}}$ be the filter on G generated by the filter base

$$\{\{1_L \mid L\in\underline{\underline{L}}_o , L\subset L'\} \mid L'\in\underline{\underline{L}}_o\} .$$

Since the intersection of any sequence in $\underline{\underline{L}}_o$ belongs to $\underline{\underline{L}}_o$, $\underline{\underline{F}}$ is a δ-filter. By the 2nd step $(\psi(1_{L_n}))_{n\in\mathbb{N}}$ is a Σ-sequence in F for any decreasing sequence $(L_n)_{n\in\mathbb{N}}$ in $\underline{\underline{L}}_o$ and therefore (Lemma 4.1.3) $\psi(\underline{\underline{F}})$ is a Cauchy filter. Hence $\psi(\underline{\underline{F}})$ is a bounded Cauchy δ-filter on F and therefore it converges in F to an x . We have

$$\langle x,x'\rangle = \lim_{L,\underline{\underline{F}}} \langle\mu(L), x'\rangle = \lim_{L,\underline{\underline{F}}} x'\circ\mu(L) = x'\circ\mu(K) =$$

$$= \langle\mu(K),x'\rangle$$

for any $x'\in E'$ and therefore $\mu(K) = x\in F$.

4th step : $\mu(A)\in F$ for any σ-compact set A of X.

Let $(K_n)_{n\in\mathbb{N}}$ be an increasing sequence of compact sets of X with union A. By the 2nd and 3rd step $(\mu(K_n))_{n\in\mathbb{N}}$ is a Σ-sequence and therefore a convergent sequence in F. We get immediately $\mu(A)\in F$.

5th step : $\mu(A)\in F$ for any $A\in\underline{\underline{R}}$.

Let $\underline{\underline{A}}$ be the set of σ-compact subsets of A and let $\underline{\underline{F}}$ be the filter on G generated by the filter base

$$\{\{1_B \mid B\in\underline{\underline{A}}\ ,\ B\supset C\} \mid C\in\underline{\underline{A}}\}\ .$$

Since the union of any sequence in $\underline{\underline{A}}$ belongs to $\underline{\underline{A}}$, $\underline{\underline{F}}$ is a δ-filter. By the 2nd and 4th step $(\psi(1_{A_n}))_{n\in\mathbb{N}}$ is a Σ-sequence in F for any increasing sequence $(A_n)_{n\in\mathbb{N}}$ in $\underline{\underline{A}}$ and therefore (Lemma 4.1.3) $\psi(\underline{\underline{F}})$ is a Cauchy filter. Hence $\psi(\underline{\underline{F}})$ is a bounded Cauchy δ-filter on F and therefore it converges in F to an x. We get immediately $\mu(A) = x\in F$.

6th step : $\int\xi d\mu\in F$ for any $\xi\in M_b^{\pi}$.

The assertion follows immediately from the 5th step and from the last assertion of Theorem 4.2.11.

b) Let now X be anbitrary. Let K be a compact set of X which is contained in a compact G_δ-set L of X. Then the compact G_δ-sets of L are exactly the compact G_δ-sets of X contained in L and therefore $\mu(M)\in F$ for any compact G_δ-set M of L. By a) we get $\mu(K)\in F$. Hence, by the hypothesis, the value of μ at any compact set of X belongs to F.

Let $\xi\in F$. There exists a compact set K of X such that $\xi = \xi i_K$. By the above considerations, by a), and by Corollary 4.2.16 d) we get

$$\int\xi d\mu = \int\xi i_K d\mu\in F\ .\ \square$$

<u>Corollary 4.4.2</u> *Let $\lambda\in M$, let Y be a compact space, and let f be*

a bounded real function on $X\times Y$ *such that* $f(\cdot,y)\in L^1_{loc}(\lambda)$ *for any* $y\in Y$. *If the map*

$$Y \longrightarrow \mathbb{R} , \ y \longmapsto \int_A f(x,y)\,d\lambda(x)$$

is continuous for any $A\in\underline{\underline{L}}$ *then it is continuous for any* $A\in\underline{\underline{R}}$.

Let us denote by E the vector space of Borel bounded real functions on Y endowed with the coarsest topology for which the map

$$E \longrightarrow \mathbb{R} , \ g \longmapsto \int g\,d\nu$$

are continuous for each Radon real measure ν on Y . We denote by F the subspace of continuous functions of E. The Mackey topology of F is the topology generated by the supremum norm and it is therefore complete.

Let $A\in\underline{\underline{R}}$. There exists an increasing sequence $(K_n)_{n\in\mathbb{N}}$ of compact subsets of A such that $A\setminus B$ is a λ-null set, where $B := \bigcup_{n\in\mathbb{N}} K_n$. Then

$$\int_A f(x,y)\,d\lambda(x) = \int_B f(x,y)\,d\lambda(x) = \lim_{n\to\infty} \int_{K_n} f(x,y)\,d\lambda(x)$$

for any $y\in Y$. We deduce that the function

$$Y \longrightarrow R , \ y \longmapsto \int_A f(x,y)\,d\lambda(x)$$

belongs to E .

We denote for any $A\in\underline{\underline{R}}$ by $\mu(A)$ the above element of E . Then $\mu\in M(E)$. By Theorem 4.4.1 $\mu(\underline{\underline{R}})\subset F$. □

<u>*Corollary 4.4.3*</u> *Let* $\mu\in M(E'^{*}_{E'})$ *and let* E *be* δ*-quasicomplete and* Σ*-complete. We assume that* $\mu(L)\in E$ *for any* $L\in\underline{\underline{L}}$. *Then* $\mu(\underline{\underline{R}})\subset E$.

E endowed with its weak topology may be considered as a subspace of $E'^{*}_{E'}$. Hence the Mackey topology of E for the topology induced by E'^{*}_{E} is finer than the initial topology of E , in particular it is δ-quasicomplete and Σ-complete. By Theorem 4.4.1 $\mu(\underline{\underline{R}})\subset E$. □

Corollary 4.4.4 *Let* $\underline{\underline{S}}$ *be a ring of sets contained in* $\underline{\underline{R}}$ *and containing* $\underline{\underline{L}}$ *, and let* ν *be an* E*-valued measure on* $\underline{\underline{S}}$ *. If* E *is* δ*-quasicomplete and* Σ*-complete then there exists a unique* $\mu \in M(E)$ *such that* $\mu|\underline{\underline{S}} = \nu$ *.*

For any $x' \in E'$ there exists a unique $\lambda_{x'} \in M$ such that $\lambda_{x'}|\underline{\underline{S}} = x' \circ \nu$. For any $A \in \underline{\underline{R}}$ we set

$$\lambda(A) : E' \longrightarrow \mathbb{R} , \; x' \longmapsto \lambda_{x'}(A) .$$

Then $\lambda \in M(E'^{*}_{E'})$. By Corollary 4.4.3 $\lambda(\underline{\underline{R}}) \subset E$. We set

$$\mu : \underline{\underline{R}} \longrightarrow E , \; A \longmapsto \lambda(A) .$$

μ possesses the required properties. The unicity is trivial. □

Corollary 4.4.5 *Let* $(\mu_n)_{n\in\mathbb{N}}$ *be a sequence in* $M(E)$ *such that* $\{\mu_n(A) | n\in\mathbb{N}\}$ *is bounded for any* $A\in\underline{\underline{R}}$ *and such that* $(\mu_n(L))$ *converges to* 0 *for any* $L\in\underline{\underline{L}}$ *. Let further* $\xi \in M^{\pi}$ *such that there exists* $A\in\underline{\underline{R}}$ *with* $\xi = \xi 1_A$ *and such that* $\int \xi d\mu_n \in E$ *for any* $n\in\mathbb{N}$ *. Then* $(\int \xi d\mu_n)_{n\in\mathbb{N}}$ *converges to* 0 *. In particular* $(\mu_n(A))_{n\in\mathbb{N}}$ *converges to* 0 *for any* $A\in\underline{\underline{R}}$ *.*

We may assume E complete. We use the notations of Lemma 4.2.23 and set

$$\mu : \underline{\underline{R}} \longrightarrow \ell^{\infty}(E) , \; A \longmapsto (\mu_n(A))_{n\in\mathbb{N}} .$$

Let F be the space $\ell^{\infty}(E)$ endowed with the $\sigma(\ell^{\infty}(E), \ell^{1}(E))$-topology and let G be the vector space $c_o(E)$ endowed with the topology induced by F . By Lemma 4.2.23 f) $\mu \in M(F)$ and by Lemma 4.2.23 a), b), e) G endowed with its Mackey topology is complete. By Theorem 4.4.1 $(\int \xi d\mu_n)_{n\in\mathbb{N}}$ converges to 0 . □

Lemma 4.4.6 (Choquet) *Let* X *be a locally compact space and let* F *be a solid subspace of the vector lattice of continuous real functions on* X *such that:*

a) any continuous real function on X *with compact carrier belongs to* F ;

β) *for any* $f \in F$ *there exists* $g \in F$ *such that the set* $\{n|f|>|g|\}$ *is relatively compact for any* $n \in \mathbb{N}$.

Let further φ *be a linear form on* F *which is bounded on the sets of the form* $\{g \in F \mid |g| \leqslant |f|\}$ *for any* $f \in F$. *Then there exists a unique* $\lambda \in M$ *such that* $f \in L^1(\lambda)$ *and*

$$\varphi(f) = \int f d\lambda$$

for any $f \in F$.

The unicity follows immediately from α) .

Let us denote by K the vector lattice of continuous real functions on X with compact carrier. There exists $\lambda \in M$ such that

$$\varphi'(f) = \int f d\lambda$$

for any $f \in K$. Let $f \in F$. By β) there exists $g \in F$ such that $\{n|f|>|g|\}$ is relatively compact for any $n \in \mathbb{N}$. We set

$$\alpha := \sup \{ |\varphi(h)| \mid h \in F \, , \, |h| \leqslant |g| \} \, .$$

We may construct inductively an increasing sequence $(f_n)_{n \in \mathbb{N}}$ in K such that

$$1_{\{n|f|>|g|\}} \leqslant f_n \leqslant 1$$

for any $n \in \mathbb{N}$. We have

$$|f|(f_n - f_m) \leqslant \frac{1}{m}|g|$$

and therefore

$$\int |f|(f_n - f_m) d\lambda = \varphi(|f|(f_n - f_m)) \leqslant \frac{\alpha}{m}$$

for any $m, n \in \mathbb{N}$ with $m \leqslant n$. Since $|f| = \bigvee_{m \in \mathbb{N}} |f| f_n$ we get $f \in L^1(\lambda)$ and

$$|\int f d\lambda - \int f f_n d\lambda| \leq \frac{\alpha}{n}$$

for any $n \in \mathbb{N}$. We have

$$|f - ff_n| \leq \frac{1}{n} |g|$$

and therefore

$$|\int f d\lambda - \varphi(f)| \leq |\int f d\lambda - \int f f_n d\lambda| + |\varphi(ff_n) - \varphi(f)| \leq$$

$$\leq \frac{\alpha}{n} + \frac{\alpha}{n} = \frac{2\alpha}{n}$$

for any $n \in \mathbb{N}$. We get

$$\varphi(f) = \int f d\lambda \ . \ \square$$

<u>*Theorem 4.4.7*</u> *Let* X *be a locally compact space and let* F *be a solid subspace of the vector lattice of continuous real functions on* X *such that:* α) *any continuous real function on* X *with compact carrier belongs to* F ; β) *for any* $f \in F$ *there exists* $g \in F$ *such that the set* $\{n|f|>|g|\}$ *is relatively compact for any* $n \in \mathbb{N}$. *Let further* G *be the solid subspace of* M_{c}^{π} *generated by* $\{\dot{f}|f \in F\}$, *let* N *be the fundamental solid subspace*

$$\{\lambda \in M | F \subset L^{1}(\lambda)\}$$

of M, *and let* φ *be a linear map* $F \longrightarrow E$. *We consider the following assertions:*

1) the set

$$\{\varphi(g) | g \in F \ , \ |g| \leq |f|\}$$

is weakly relatively compact for any $f \in F$;

2) there exists $\mu \in N(E)$ *such that*

$$\varphi(f) = \int f d\mu$$

for any $f \in F$ *and* $\int \xi d\mu \in E$ *for any* $\xi \in G$;

3) there exists $\mu \in N(E)$ *such that*

$$\varphi(f) = \int f d\mu$$

for any $f \in F$ *;*

4) there exists $\mu \in N(E'^{*}_{E'})$ *such that (by identifying* E *with a subspace of* E'*)

$$\varphi(f) = \int f d\mu$$

for any $f \in F$ *and* $\mu(K) \in E$ *for any compact* G_δ*-set* K *of* X .

Then:

a) $1 \Longleftrightarrow 2 \Longrightarrow 3 \Longrightarrow 4$ *;*

b) if E *is* δ*-quasicompact and* Σ*-complete then* $4 \Longrightarrow 2$ *;*

c) the set of continuous real functions on X *with compact carrier and the set of continuous real functions on* X *vanishing at* ∞ *possess the above properties* α) *and* β)*;*

d) if X *is* σ*-compact then the set* C *of all continuous real functions on* X *possesses the above properties* α) *and* β) *; if* F = C *then* 2) *implies* $\{\int \xi d\mu \mid \xi \in A\}$ *is a precompact set of* E *for any weakly pseudo-compact set* A *of* (G,H) *where* $H := \{\dot{i}_A \mid A \in \underline{\underline{R}}\}$.

a) We have

$$(\lambda|\dot{F})^{+} = \lambda^{+}|\dot{F}$$

for any $\lambda \in N$, where $\dot{F} = \{\dot{f} \mid f \in F\}$.

$1 \Longrightarrow 2$. Let $x' \in E'$. By 1) $x' \circ \varphi$ is bounded on the sets of the form $\{g \in F \mid |g| \leqslant |f|\}$ for any $f \in F$. By Lemma 4.4.6 there exists $\lambda_{x'} \in N$ with

$$x' \circ \varphi(f) = \int f d\lambda_{x'}$$

for any $f \in F$. It follows φ is continuous with respect to the Mackey topology $\tau(F,N)$ and 2) follows immediately from Proposition 4.3.9 a).

$2 \Longrightarrow 1$ follows from Proposition 4.3.9 a) .

$2 \Rightarrow 3 \Rightarrow 4$ is trivial.

b) By Corollary 4.4.3 $\mu(\underline{\underline{R}}) \subset E$ and so we may consider $\mu \in N(E)$. Let $\xi \in G$. There exists $f \in F$ such that $|\xi| \leq |\dot{f}|$. There exists further $g \in F$ such that the set $\{n|f| > |g|\}$ is relatively compact for any $n \in \mathbb{N}$. By Proposition 4.3.8 $\int \xi d\mu \in E$.

c) If F is the set of continuous real functions on X with compact carrier then $\{n|f| > |f|\}$ is relatively compact for any $n \in \mathbb{N}$ and $\alpha)$ & $\beta)$ is fulfilled. Assume now F is the set of continuous real functions on X vanishing at ∞ and let $f \in F$. Then $|f|^{1/2} \in F$ and from

$$\{n|f| > |f|^{1/2}\} \subset \{|f| > \frac{1}{n^2}\}$$

we see that $\{n|f| > |f|^{1/2}\}$ is relatively compact for any $n \in \mathbb{N}$. Hence $\alpha)$ & $\beta)$ is fulfilled.

d) Let f be a continuous real function on X and let g be a positive continuous real function on X converging to ∞ at the infinity. Then $|f|^2 + g \in F$ and the set $\{n|f| > |f|^2 + g\}$ is relatively compact for any $n \in \mathbb{N}$. Hence $\alpha)$ & $\beta)$ is fulfilled. The last assertion follows from Theorem 4.3.5. □

Remark. R.G. Bartle, N. Dunford and J. Schwartz proved the equivalence $1 \Leftrightarrow 3$ for X compact and E a Banach space ([2] Theorem 3.2). D.R. Lewis proved the implication $1 \Rightarrow 2$ for X compact and the implication $2 \Rightarrow 1$ for X compact and E complete, the elements ξ in 2) being replaced in both implications by bounded Borel functions on X ([21] Theorem 3.1).

Theorem 4.4.8 *Let X be a locally compact space, let F be a solid subspace of the vector lattice of continuous real functions on X containing any such function having a compact carrier, let G be the solid subspace of M_c^π generated by $\{\dot{f} | f \in F\}$, let N be the fundamental solid subspace*

$$\{\lambda \in M | F \subset L^1(\lambda)\}$$

of M, and let φ be a linear map $F \to E$. We consider the following assertions :

1) $\{\varphi(g) \mid g\in F\ ,\ |g|\leq|f|\}$ *is weakly relatively compact for any* $f\in F$;

2) *there exists* $\mu\in N(E)$ *such that*

$$\varphi(f) = \int f d\mu$$

for any $f\in F$ *and* $\int \xi d\mu \in E$ *for any* $\xi\in G$;

3) *there exists* $\mu\in N(E)$ *such that*

$$\varphi(f) = \int f d\mu$$

for any $f\in F$;

4) α) *there exists* $\mu\in N(E'^{*}_{E'})$ *such that* (by *identifying* E *with a subspace of* E'^{*})

$$\varphi(f) = \int f d\mu$$

for any $f\in F$ *and* β) $\mu(K)\in E$ *for any compact* G_δ*-set* K *of* X ;

5) *for any lower directed subset* A *of* F *such that*

$$\inf_{f\in A} f(x) = 0$$

for any $x\in X$, $\varphi(\underline{\underline{F}})$ *converges weakly to* 0, *where* $\underline{\underline{F}}$ *denotes the filter on* F *generated by the filter base*

$$\{\{g\in A \mid g\leq f\} \mid f\in F\} .$$

Then:

a) $1\ \&\ 5 \Leftrightarrow 2 \Rightarrow 3 \Rightarrow 4 \Rightarrow 5$;

b) *if* E *is* δ*-complete and* Σ*-complete then* $4 \Rightarrow 3$;

c) *if* E *is* δ*-complete and* Σ*-complete and if for any* $\xi\in G$ *there exists* $\eta\in\hat{L}^1(\mu)$ *such that* $(n|\xi|-|\eta|)\vee 0$ *belongs to the solid subspace of* M^{π}_{c} *generated by* $\{i_A \mid A\in\underline{\underline{R}}\}$ *for any* $n\in\mathbb{N}$ *then* $4\Rightarrow 2$; *if moreover* E *is weakly* Σ*-complete then* $4\alpha \Rightarrow 2$.

a) $1\ \&\ 5 \Rightarrow 2$. Let us denote by K the set of functions of F with compact carrier and by ψ the restriction of φ to K . By

Theorem 4.4.7 a), c) there exists $\mu \in M(E)$ such that

$$\psi(f) = \int f d\mu$$

for any $f \in K$.

Let $f \in F$. We have

$$\sup\{\int g d(x' \circ \mu) \mid g \in K \ , \ |g| \leqslant |f|\} =$$

$$= \sup\{\langle \psi(g) \ , \ x' \rangle \mid g \in K \ , \ |g| \leqslant |f|\} \leqslant$$

$$\leqslant \sup\{\langle x, x' \rangle \mid x \in \{\varphi(g) \mid g \in F \ , \ |g| \leqslant |f|\}\} < \infty$$

and therefore $f \in L^1(x' \circ \mu)$ for any $x' \in E'$. Hence $\mu \in N(E)$.

Let $f \in F_+$. We set

$$A := \{g \in K_+ \mid g \leqslant f\}$$

and denote by $\underline{\underline{F}}$ the filter on K generated by the filter base

$$\{\{h \in A \mid g \leqslant h\} \mid g \in A\} \ .$$

We have

$$\int f d(x' \circ \mu) = \lim_{g, \underline{\underline{F}}} \int g d(x' \circ \mu) = \lim_{g, \underline{\underline{F}}} \langle \varphi(g) \ , \ x' \rangle$$

for any $x' \in E'$. $\{f-g \mid g \in A\}$ is a lower directed subset of F such that

$$\inf_{g \in A} (f-g)(x) = 0$$

for any $x \in X$. By 5) $\varphi(\underline{\underline{F}})$ converges weakly to $\varphi(f)$ and therefore

$$\varphi(f) = \int f d\mu \ .$$

Hence φ is the adjoint of the map

$$E' \longrightarrow N \, , \; x' \longmapsto x' \circ \mu$$

and therefore it is continuous with respect to the Mackey topology $\tau(F, N)$. By Proposition 4.3.9 a) $\int \xi d\mu \in E$ for any ξ belonging to the solid subspace of M_c^π generated by $\{\dot{f} \mid f \in F\}$.

$2 \Rightarrow 3 \Rightarrow 4$ is trivial.

$4 \Rightarrow 5$. φ is the adjoint of the map

$$E' \longrightarrow N \, , \; x' \longmapsto x' \circ \mu$$

and therefore it is continuous with respect to the weak topologies $\sigma(F,N)$ and $\sigma(E,E')$. $\underline{\underline{F}}$ converges to 0 with respect to the $\sigma(F,N)$-topology and therefore $\varphi(\underline{\underline{F}})$ converges weakly to 0 .

$2 \Rightarrow 1$ & 5 . 1) follows from Theorem 4.2.20 b) and 5) from the above proved implications.

b) By Corollary 4.4.3 $\mu(\underline{\underline{R}}) \subset E$ and therefore we may consider $\mu \in N(E)$.

c) By b) it is sufficient to prove $3 \Rightarrow 2$ and this follows from Proposition 4.3.8. The last assertion follows from Proposition 4.2.18.□

Remarks 1. If E is quasicomplete and weakly Σ-complete, F is K, and φ is continuous then by c) all assertions hold. This was proved by E. Thomas ([30] § 5).

2. The implication $3 \Rightarrow 2$ does not hold even if E is complete, X metrizable and countable, and F the set of continuous bounded real functions on X . We take as E the vector space of bounded real functions g on $\mathbb{N} \times \mathbb{N}$ such that

$$\lim_{n\to\infty} \sup_{m\in\mathbb{N}} |g(m,n)| = 0 \, ,$$

and such that the set of $n \in \mathbb{N}$ for which the sequence $(g(m,n))_{m\in\mathbb{N}}$ does not converge is finite. For any sequence $\alpha := (\alpha_n)_{n\in\mathbb{N}}$ of real numbers we denote by p_α the seminorm on E

$$E \longrightarrow \mathbb{R}_+ \, , \; g \longrightarrow \sup_{m,n\in\mathbb{N}} |g(m,n)| \; +$$

$$+ \sum_{n\in\mathbb{N}} |\alpha_n| \, \left|\limsup_{m\to\infty} g(m,n) - \liminf_{m\to\infty} g(m,n)\right| .$$

We endow E with the topology generated by these seminorms. E is complete. We take as X the subspace

$$\{(0,n)\,|\,n\in\mathbb{N}\}\cup\{(\tfrac{1}{m},n)\,|\,m\in\mathbb{N}\setminus\{0\}\ ,\ n\in\mathbb{N}\}$$

of $\mathbb{R}^2$. We denote for any bounded real function f on X by ψf the map

$$N\times N\longrightarrow\mathbb{R}\ ,\ (m,n)\longmapsto\begin{cases}\frac{1}{n}f(\frac{1}{m},n) & \text{if } m\neq 0\\ \frac{1}{n}f(0,n) & \text{if } m=0\ .\end{cases}$$

We set

$$\varphi : F\longrightarrow E\ ,\ f\longmapsto\psi f\ ,$$

$$\mu : \underline{\underline{R}}\longrightarrow E\ ,\ A\longmapsto\psi 1_A\ .$$

φ is a linear map, $\mu\in M(E)$, and

$$\varphi(f)=\int f d\mu$$

for any $f\in F$. So 3) is fulfilled by 2) fails since

$$\int 1_{A\times N}d\mu\notin E\ ,$$

where $A:=\{2n+1\,|\,n\in\mathbb{N}\}$.

<u>*Proposition 4.4.9*</u> *Let X be a completely regular space, let C_b be the vector space of continuous bounded real functions on X , and let φ be a linear map $C_b\longrightarrow E$. Then the following assertions are equivalent:*

a) *φ is continuous with respect to the Mackey topology $\tau(C_b,M_b)$ and $\{\varphi(f)\,|\,f\in C_b\ ,\ |f|\leq 1\}$ is a weakly relatively compact set of E ;*

b) *there exists $\mu\in M_b(E)$ such that*

$$\varphi(f)=\int f d\mu$$

for any $f\in C_b$ and $\int\xi d\mu\in E$ for any $\xi\in M_b^{\pi}$.

We have

$$(\lambda|\dot{C}_b)^+ = \lambda^+|\dot{C}_b$$

for any $\lambda \in M_b$, where $\dot{C}_b := \{\dot{f}|f\in C_b\}$ and the assertion follows immediately from Proposition 4.3.9 a) . □

Remark. A more general result was proved by Katsaras ([20] Theorem 3).

Proposition 4.4.10 *Let* X *be a locally compact paracompact space and let* $\mu \in M(E)$. *We denote by* K *the vector lattice of continuous real functions on* X *with compact carrier and endow* K *and* M^π *with the order topology. We assume* $\int \xi d\mu \in E$ *for any* $\xi \in K$ *(any* $\xi \in M^\pi$*). Then:*

a) $\{\int \xi d\mu | \xi \in A\}$ *is a precompact set of* E *for any weakly pseudo-compact set* A *of* K *(of* M^π*) ;*

b) $\{\int \xi d\mu | \xi \in A\}$ *is a compact set of* E *for any weakly compact set* A *of* K *(of* M^π*) .*

The assertions concerning M^π follow immediately from Corollary 4.3.11. In order to prove the assertions for K we may assume E complete. Since X is locally compact and paracompact, for any $\xi \in M^\pi$ there exists $A \in \underline{\underline{R}}$ such that $\xi = \xi \dot{1}_A$. By Corollary 4.2.17 a) $\int \xi d\mu \in E$ for any $\xi \in M^\pi$. The map

$$K \longrightarrow M^\pi , \ f \longmapsto \dot{f}$$

being continuous, it is continuous for the weak topologies and therefore the image of any weakly pseudo-compact (weakly compact) set of K is a weakly pseudo-compact (weakly compact) set of M^π . The assertions concerning K follow from the assertions concerning M^π and the above remark. □

Theorem 4.4.11 *Let* X *be a locally compact paracompact space, let* C *be the vector space of continuous real functions on* X *endowed with the topology of compact convergence, let* $F := \{\dot{1}_A | A \in \underline{\underline{R}}_{\Gamma}\}$, *and let* $\mu \in M(E)$. *We assume* $\dot{f} \in \hat{L}^1(\mu)$ *and* $\int f d\mu \in E$ *for any* $f \in C$ *(*$\xi \in \hat{L}^1_{F}(\mu)$ *and* $\int \xi d\mu \in E$ *for any* $\xi \in M^\pi_C$*) . Then:*

a) there exists $A \in \underline{\underline{R}}$ *such that* $\mu(B) = 0$ *for any* $B \in \underline{\underline{R}}$ *with* $A \cap B = \emptyset$;

b) $\{\int \xi d\mu \,|\, \xi \in A\}$ *is a precompact set of* E *for any weakly pseudo-compact set* A *of* $C(\text{of } (M_c^{\pi}, F))$;

c) $\{\int \xi d\mu \,|\, \xi \in A\}$ *is a compact set of* E *for any weakly compact set* A *of* $C(\text{of } (M_c^{\pi}, F))$.

a) If a) does not hold then there exists a sequence $(A_n)_{n \in \mathbb{N}}$ in $\underline{\underline{R}}$, such that $\mu(A_n) \neq 0$ for any $n \in \mathbb{N}$ and such that $\{n \in \mathbb{N} \,|\, A_n \cap K \neq 0\}$ is finite for any compact set K of X . It is easy to construct an $f \in C$ such that $\dot{f} \notin \hat{\mathcal{L}}^1(\mu)$.

b & c. We may assume E complete. Let $\xi \in M_c^{\pi}$. By Corollary 4.2.17a) $\int \xi i_A d\mu \in E$. We deduce by a) $\xi \in \hat{\mathcal{L}}^1(\mu)$ and

$$\int \xi d\mu = \int \xi i_A d\mu \in E \, .$$

Since the map

$$C \longrightarrow (M_{\pi}^{c}, F) \, , \; f \longmapsto \dot{f}$$

is continuous it is sufficient to prove the assertions for (M_{π}^{c}, F) only. The map

$$M_c^{\pi} \longrightarrow E \, , \; \xi \longmapsto \int \xi d\mu$$

being obviously continuous for the $\tau(M_c^{\pi}, M_c)$-topology the assertions follow from Corollary 3.8.6. □

Theorem 4.4.12 *Let* X *be a locally compact paracompact space, let* $F := \{i_A \,|\, A \in \underline{\underline{R}}\}$, *and let* $(\mu_n)_{n \in \mathbb{N}}$ *be a sequence in* M(E) *such that* $(\mu_n(A))_{n \in \mathbb{N}}$ *is bounded for any* $A \in \underline{\underline{R}}$ *and such that* $(\mu_n(L))_{n \in \mathbb{N}}$ *converges to* 0 *for any compact* G_δ*-set of* X . *We denote by* G *one of the following locally convex space :* (M_c^{π}, F), M^{π} *endowed with the order topology, the vector space of continuous real functions on* X *endowed with the topology of compact convergence, the vector lattice of continuous real functions on* X *with compact carrier endowed with the order topology. If* $\xi \in \hat{\mathcal{L}}^1(\mu_n)$ *and* $\int \xi d\mu_n \in E$ *for any* $\xi \in G$ *and for any* $n \in \mathbb{N}$ *then* $(\int \xi d\mu_n)_{n \in \mathbb{N}}$ *converges to* 0 *uniformly in* ξ *for* ξ *belonging to a weakly pseudo-compact set of* G .

We may assume E complete. By Corollary 4.4.5 $(\mu_n(A))_{n\in\mathbb{N}}$ converges to 0 for any $A\in\underline{\underline{R}}$. We use the notations of Lemma 4.2.23 and set

$$\mu : \underline{\underline{R}} \longrightarrow c_o(E) \text{ , } A \longmapsto (\mu_n(A))_{n\in\mathbb{N}}$$

By Lemma 4.2.23 g) $\mu\in M(c_o(E))$ and by Lemma 4.2.23 a), b) $c_o(E)$ is complete, where $c_o(E)$ is endowed with the topology induced by $\ell^\infty(E)$. Since X is locally compact and paracompact for any $\xi\in G$ there exists $A\in\underline{\underline{R}}$ such that $\xi \dot{i}_A\in\hat{L}^1(\mu)$ and

$$\int \xi d\mu = \int \xi \dot{i}_A d\mu$$

(Theorem 4.4.11 a)). By Corollary 4.2.17 a) $\int\xi d\mu\in c_o(E)$ for any $\xi\in M^\pi$.

Let A be a weakly pseudo-compact set of G . By Proposition 4.4.10 a) and Theorem 4.4.11 b)

$$\{\int \xi d\mu \mid \xi\in A\}$$

is a precompact set of $c_o(E)$. Let p be a continuous seminorm on E. Then

$$\lim_{n\to\infty} p(x_n) = 0$$

uniformly in $(x_n)_{n\in\mathbb{N}}$ for $(x_n)_{n\in\mathbb{N}}$ belonging to a precompact set of $c_o(E)$ and therefore $(\int\xi d\mu_n)_{n\in\mathbb{N}}$ converges to 0 uniformly on A . □

5. Topologies on the spaces of vector measures

Definition 4.5.1 *Let* N *be a fundamental solid subspace of* M. *We denote for any* $\xi\in N^\pi$ *and for any equicontinuous set* A' *of* E' *by* $q_{\xi,A'}$ *the seminorm on* $N(E)$

$$N(E) \longrightarrow \mathbb{R} \text{ , } \mu \longmapsto \sup_{x'\in A'} \int |\xi| d|x'\circ\mu|$$

(Proposition 4.3.6 a)): *for any subset* $\mathcal{F}$ *of* N^{π} *and for any subspace* F *of* $N(E)$ *we denote by* $(F,\mathcal{F})$ *the space* F *endowed with the topology generated by the family of seminorms* $(q_{\xi,A'})$ *where* ξ *runs through* $\mathcal{F}$ *and* A' *through the set of equicontinuous sets of* E' (if $\mathcal{F}$ generates N^{π} as band then $(F,\mathcal{F})$ is Hausdorff).

Theorem 4.5.2 *Let* N *be a fundamental solid subspace of* M *and let* $\mathcal{F}$ *be a solid subspace of* N^{π} *containing* $\{i_A | A\in \underline{\underline{R}}\}$ *such that* $\int \xi d\mu \in E$ *for any* $(\xi,\mu)\in \mathcal{F}\times N(E)$ *. Let further* $L(\mathcal{F},E)$ *be the vector space of linear maps of* $\mathcal{F}$ *into* E *which are continuous with respect to the Mackey topology* $\tau(\mathcal{F},N)$ *and let us endow* $L(\mathcal{F},E)$ *with the topology of uniform convergence on the order bounded sets of* $\mathcal{F}$ *. We denote for any* $\mu\in N(E)$ *by* μ' *the map*

$$\mathcal{F} \longrightarrow E\ ,\ \xi \longmapsto \int \xi d\mu\ .$$

Then:

a) $\mu'\in L(\mathcal{F},E)$ *for any* $\mu\in N(E)$ *;*

b) the map

$$(N(E),\mathcal{F}) \longrightarrow L(\mathcal{F},E),\ \mu \longmapsto \mu'$$

is an isomorphism of locally convex spaces ;

c) if E *is complete then so is* $N(E)$ *.*

a) follows from Proposition 4.3.1 c $\Longrightarrow$ a .

b) By Proposition 4.3.1 a $\Longleftrightarrow$ c the map

$$N(E) \longrightarrow L(\mathcal{F},E),\ \mu \longmapsto \mu'$$

is an isomorphism of vector spaces. Let p be a continuous seminorm on E , let

$$U := \{x\in E | p(x) \leqslant 1\}\ ,$$

let U° be the polar set of U in E', and let $\xi\in\mathcal{F}$. We have

$$\sup_{x'\in U^{\circ}} \int |\xi| d|x'\circ\mu| = \sup\{\int \eta d(x'\circ\mu) | \eta\in\mathcal{F},\ |\eta| \leqslant |\xi|,\ x'\in U^{\circ}\} =$$

$= \sup\{<\int \eta d\mu, x'> | \eta \in F, |\eta| \leq |\xi|, x' \in U^{\circ}\} =$

$= \sup \{p(\mu'(\eta)) \ | \eta \in F, |\eta| \leq |\xi|\}$

for any $\mu \in N(E)$. Hence the map

$$N(E) \longrightarrow L(F,E), \ \mu \longmapsto \mu'$$

is an isomorphism of locally convex spaces.

c) Let $\underline{\underline{F}}$ be a Cauchy filter on $L(F,E)$ and let φ be the map

$$F \longrightarrow E \ , \ \xi \longrightarrow \lim_{\psi,\underline{\underline{F}}} \psi(\xi) \ .$$

We want to show φ is order continuous. Let A be a lower directed nonempty set of F with infinum 0, let $\xi \in A$, and let U be a closed convex 0-neighbourhood in E . There exists $B \in \underline{\underline{F}}$ such that

$$\psi'(\eta) - \psi''(\eta) \in \frac{1}{2}U$$

for any $\psi', \psi'' \in B$ and $\eta \in F$ with $0 \leq \eta \leq \xi$. Let $\psi \in B$. We get

$$\psi(\eta) - \varphi(\eta) \in \frac{1}{2}U$$

for any $\eta \in F$ with $0 \leq \eta \leq \xi$. By Proposition 4.3.1 a $\Longrightarrow$ b ψ is order continuous. Hence there exists $\eta \in A$ such that $\psi(J) \in \frac{1}{2}U$ for any $J \in A$ with $J \leq \eta$. We get $\varphi(J) \in U$ for any $J \in A$ with $J \leq \xi \wedge \eta$. Hence φ is order continuous and therefore by Proposition 4.3.1 b $\Longrightarrow$ a it belongs to $L(F,E)$. It is obvious that $\underline{\underline{F}}$ converges to φ . Hence $L(F,E)$ is complete. By b) $N(E)$ is complete. □

<u>*Theorem 4.5.3*</u> *Let* N *be a fundamental solid subspace of* M, *let* F *be a solid subspace of* N^{π} *containing* $\{i_A | A \in \underline{\underline{R}}\}$ *such that* $\int \xi d\mu \in E$ *for any* $(\xi,\mu) \in F \times N(E)$, *and let* $N_o(E)$ *be the set of* $\mu \in N(E)$ *for which the set*

$$\{\int \eta d\mu \,|\, \eta \in F \ , \ |\eta| \leq |\xi|\}$$

is compact for any $\xi \in F$. *Let further* E'_1 *and* E'_2 *be the vector space* E' *endowed with the topology of uniform convergence on the weakly compact convex and compact convex sets of* E *respectively and for any* $i\in\{1,2\}$ *let* L_i *be the vector space of continuous linear maps of* E'_i *into* (N,F) *endowed with the topology of uniform convergence on the equicontinuous subsets of* E' *. We denote for any* $\mu\in N(E)$ *by* μ' *the map*

$$E' \longrightarrow N, \quad x' \longmapsto x' \circ \mu .$$

Then:

a) $\mu'\in L_1$ *for any* $\mu\in N(E)$ *;*

b) $\mu'\in L_2 \Leftrightarrow \mu\in N_o(E)$ *for any* $\mu\in N(E)$ *;*

c) the maps

$$N(E) \longrightarrow L_1, \quad \mu \longmapsto \mu',$$

$$N_o(E) \longrightarrow L_2, \quad \mu \longmapsto \mu'$$

are isomorphims of locally convex spaces.

a) follows immediately from Proposition 3.1.8 and Theorem 4.5.2 a).

b) Assume first $\mu\in N_o(E)$. Let $\xi\in F$. We set

$$K := \{\int \eta d\mu \mid \eta\in F, \ |\eta| \leq |\xi|\}$$

and denote by K^o its polar set in E' . Then

$$q_\xi(x'\circ\mu) = \int|\xi|d|x'\circ\mu| = \sup\{\int \eta d(x'\circ\mu) \mid \eta\in F, \ |\eta| \leq |\xi|\} =$$

$$= \sup\{\langle\int\eta d\mu, x'\rangle \mid \eta\in F, \ |\eta| \leq |\xi|\} \leq 1$$

for any $x'\in K^o$. Since K is a compact convex set of E we get $\mu'\in L_2$.

Assume now $\mu\in L_2$. Let $\xi\in F$. There exists a compact convex set K of E such that

$$q_\xi(x'\circ\mu) \leq 1$$

for any $x'\in K^{o}$, where K^{o} denotes the polar set of K in E'. We get

$$|<\int \eta d\mu , x'>| \leqslant \int |\xi| d|x'\circ\mu| \leqslant 1$$

for any $x'\in K^{o}$ and for any $\eta\in F$ with $|\eta| \leqslant |\xi|$ and therefore

$$\{\int \eta d\mu \mid \eta\in F , |\eta| \leqslant |\xi|\} \subset K .$$

Hence $\mu\in N_{o}(E)$.

c) The map

$$N(E) \longrightarrow L_{1} , \mu \longmapsto \mu'$$

is obviously linear. Let $u\in L_{1}$. By Proposition 3.4.2 F is the dual of (N,F). Let $u': F \longrightarrow E$ be the adjoint of u. By Proposition 4.3.1 a $\Longrightarrow$ c there exists $\mu\in M(E)$ such that $\xi\in\hat{L}^{1}(\mu)$ and

$$\int \xi d\mu = u'(\xi)$$

for any $\xi\in F$. We get

$$x'\circ\mu = u(x')$$

for any $x'\in E'$ and therefore $\mu\in N(E)$ and $\mu' = u$. Hence the map

$$N(E) \longrightarrow L_{1} , \mu \longmapsto \mu'$$

is surjective.

Let A' be an equicontinuous subset of E', let $\xi\in F$, and let $\mu\in N(E)$. From

$$q_{\xi,A'}(\mu) = \sup_{x'\in A'} \int |\xi| d|x'\circ\mu| = \sup_{x'\in A'} \int |\xi| d|\mu'(x')| =$$

$$= \sup_{x'\in A'} q_{\xi}(\mu'(x'))$$

it follows that the map

$$N(E) \longrightarrow L_1 \ , \ \mu \longmapsto \mu'$$

is an isomorphism of locally convex spaces. We deduce further by b) that the map

$$N_o(E) \longrightarrow L_2 \ , \ \mu \longmapsto \mu'$$

is an isomorphism of locally convex spaces too. □

Lemma 4.5.4 *Let* E, F *be locally convex spaces, let* G *be the vector space of linear maps* $z : F' \longrightarrow E$ *continuous with respect to the* $\tau(F',F)$*-topology and such that* $z(B')$ *is a precompact set for any equicontinuous subset* B' *of* F'. *We endow* G *with the topology of uniform convergence on the equicontinuous subset of* F' . *Then:*

a) for any equicontinuous subset B' *of* F' *and for any precompact set* C *of* G *the set* $\bigcup_{z\in C} z(B')$ *is precompact;*

b) for any $z\in G$ *its adjoint map* $z' : E' \longrightarrow F$ *is continuous with respect to the topology of precompact convergence on* E' ;

c) for any equicontinuous subset A' *of* E' *and for any precompact set* C *of* G *the set* $\bigcup_{z\in C} z'(A')$ *is precompact, where* $z' : E' \longrightarrow F$ *denotes the adjoint of* z *for any* $z\in C$;

d) let u *be a continuous linear map of* E *into itself, let* v *be a continuous linear map of* F *into itself, and let* v' *be the adjoint map of* v *; then* $u\circ z\circ v' \in G$ *for any* $z\in G$ *and the map*

$$w : G \longrightarrow G \ , \ z \longmapsto u\circ z\circ v'$$

is continuous and linear; if u(E) *and* v(F) *have finite dimensions then* w(G) *has a finite dimension too;*

e) if E *and* F *possess the approximation property then so does* G

a) Let U be a convex 0-neighbourhood in E . We set

$$W := \{z\in G \mid z(B') \subset \tfrac{1}{2} U\} \ .$$

W is a 0-neighbourhood in G. Hence there exists a finite subset C_o

of G such that $C \subset C_o + W$. The set $\bigcup_{z \in C_o} z(B')$ being precompact there exists a finite subset A_o of E such that

$$\bigcup_{z \in C_o} z(B') \subset A_o + \frac{1}{2} U .$$

Let $z \in C$ and let $y' \in B'$. There exists $z \in C_o$ with $z - z_o \in W$. We get

$$z(y') - z_o(y') = (z - z_o)(y') \in \frac{1}{2} U$$

and therefore

$$z(y') \in A_o + U .$$

Since z and y' are arbitrary we deduce

$$\bigcup_{z \in C} z(B') \subset A_o + U .$$

Hence $\bigcup_{z \in C} z(B')$ is precompact.

b) Let V be a closed convex 0-neighbourhood in F and let V^o be its polar set in F'. Then $z(V^o)$ is a precompact set of E and therefore its polar set $z(V^o)^o$ in E' is a 0-neighbourhood in E' for the topology of precompact convergence in E'. Let $x' \in z(V^o)^o$ and let $y' \in V^o$. We have

$$\langle z'(x'), y' \rangle = \langle x', z(y') \rangle \leqslant 1 .$$

Since y' is arbitrary we get $z'(x') \in V$. Hence $z(V^o)^o \subset \overset{-1}{z}{}'(V)$. $\overset{-1}{z}{}'(V)$ is therefore a 0-neighbourhood in E' for the topology of precompact convergence on E' and z' is continuous with respect to this topology.

c) Let V be a closed convex 0-neighbourhood in F. We denote by V^o the polar set of V in F' and by A'^c the polar set of A' in E. We set

$$W := \{z \in G \mid z(V^o) \subset \frac{1}{2} A'^o\} .$$

W is a 0-neighbourhood in G. Hence there exists a finite subset C_o

of C such that $C \subset C_o + W$. The set A' is precompact with respect to the $\sigma(E',E)$-topology. Being equicontinuous it is precompact with respect to the topology of precompact convergence on E'. By b) the set $\bigcup_{z \in C_o} z'(A')$ is a precompact set of F. Hence there exists a finite subset B_o of F such that

$$\bigcup_{z \in C_o} z'(A') \subset B_o + \frac{1}{2} V .$$

Let $z \in C$ and let $x' \in A'$. There exists $z_o \in C$ such that $z - z_o \in W$. We have

$$\langle z'(x') - z_o'(x'), y' \rangle =$$

$$= \langle x', z(y') - z_o(y') \rangle = \langle x', (z-z_o)(y') \rangle \leqslant \frac{1}{2}$$

for any $y' \in V^o$ and therefore $z'(x') - z_o'(x') \in \frac{1}{2} V$. We get

$$z'(x') \in B_o + V .$$

Since x' and z are arbitrary we deduce

$$\bigcup_{z \in C} z'(A') \subset B_o + V .$$

Hence $\bigcup_{z \in C} z'(A')$ is a precompact set of F.

d) v' is continuous with respect to the $\tau(F',F)$-topology and therefore $u \circ z \circ v'$ is continuous with respect to this topology too. Let B' be an equicontinuous subset of F'. Then $v'(B')$ is equicontinous and therefore $z \circ v'(B')$ and $u \circ z \circ v'(B')$ are precompact sets of E. Hence $u \circ z \circ v' \in G$.

It is obvious that w is linear. Let W be a 0-neighbourhood in G. There exists an equicontinous subset B' of F' and a 0-neighbourhood U in E such that

$$\{ z \in G \mid z(B') \subset U \} \subset W .$$

We set

$$W_o := \{z\in G \mid z(v'(B'))\subset \bar{u}^{1}(U)\} .$$

W_o is a 0-neighbourhood in G and we have

$$(w(z))(B') = u(z(v'(B')))\subset U$$

for any $z\in W_o$. Hence $w(W_o)\subset W$ and w is continuous.

Assume now that $u(E)$ and $v(F)$ have finite dimensions. There exist a finite family $((x_\iota,x'_\iota))_{\iota\in I}$ in $E\times E'$ and a finite family $((y_\lambda,y'_\lambda))_{\lambda\in L}$ in $F\times F'$ such that

$$u(x) = \sum_{\iota\in I} \langle x,x'_\iota\rangle x_\iota ,$$

$$v(y) = \sum_{\lambda\in L} \langle y,y'_\lambda\rangle y_\lambda ,$$

for any $x\in E$ and for any $y\in F$. We get

$$v'(y') = \sum_{\lambda\in L} \langle y_\lambda,y'\rangle y'_\lambda$$

for any $y'\in F'$. Let $z\in G$. We have

$$(w(z))(y') = u\circ z\circ v'(y') = u(\sum_{\lambda\in L} \langle y_\lambda,y'\rangle z(y'_\lambda)) =$$

$$= \sum_{\lambda\in L} \langle y_\lambda,y'\rangle \sum_{\iota\in I} \langle z(y'_\lambda), x'_\iota\rangle x_\iota$$

for any $y'\in F'$ and therefore

$$w(z) = \sum_{(\iota,\lambda)\in I\times L} \langle z(y'_\lambda), x'_\iota\rangle\, x_\iota\otimes y_\lambda .$$

This shows that $w(G)$ is finite dimensional.

e) Let C be a precompact set of G and W be a 0-neighbourhood in G . We have to show that there exists a continuous linear map w of G into itself such that $w(G)$ has a finite dimension and

$$w(z)-z\in W$$

for any $z\in C$. We may assume there exists an equicontinuous subset B' of F' and a closed convex 0-neighbourhood U in E such that

$$W = \{z\in G \mid z(B')\subset U\}.$$

Let U° be the polar set of U in E' and for any $z\in C$ let $z': E' \longrightarrow F$ be the adjoint map of z . By c)

$$B := \bigcup_{z\in C} z'(U^{\circ})$$

is a precompact set of F . Since F possesses the approximation property there exists a continuous linear map v of F into itself such that $v(F)$ is finite dimensional and such that

$$vy-y\in\frac{1}{2} B'^{\circ}$$

for any $y\in B$, where B'° denotes the polar set of B' in F . Let $v': F' \longrightarrow F'$ be the adjoint map of v and let $y'\in B'$. We have

$$|\langle v'y'-y', y\rangle = \langle y',vy-y\rangle \leq\frac{1}{2}$$

for any $y\in B$ and therefore $v'y'-y'\in\frac{1}{2} B^{\circ}$, where B° denotes the polar set of B in F' .

Let $z\in C$ and let $y'\in B'$. We have

$$\langle z(v'y'-y'), x'\rangle = \langle v'y'-y', z'x'\rangle \leq\frac{1}{2}$$

for any $x'\in U^{\circ}$ and therefore

$$z\circ v'(y')-z(y') = z(v'y'-y')\in\frac{1}{2} U .$$

$v'(B')$ is an equicontinuous subset of F' . By a)

$$A := \bigcup_{z\in C} z(v'(B'))$$

is a precompact set of E . Since E possesses the approximation property there exists a continuous linear map u of E into itself such that $u(E)$ is finite dimensional and such that

$$ux-x\in\frac{1}{2}U$$

for any $x\in A$.

We set

$$w : G \longrightarrow G , \; z \longmapsto u\circ z\circ v' .$$

By d) w is a continuous linear map and $w(G)$ has a finite dimension. Let $z\in G$. We have

$$(wz-z)(y') = u\circ z\circ v'(y')-z(y') =$$

$$= (u(z\circ v'(y'))-z\circ v'(y'))+(z\circ v'(y')-z(y'))\in\frac{1}{2}U+\frac{1}{2}U = U$$

for any $y'\in B'$ and therefore

$$wz-z\in W .$$

Hence G possesses the approximation property. □

Theorem 4.5.5 *Let N be a fundamental solid subspace of M and let F be a solid subspace of N^{π} such that $\int\xi d\mu\in E$ for any $(\xi,\mu)\in F\times N(E)$. We denote by $N_f(E)$ the set of $\mu\in N(E)$ for which $\{\int\xi d\mu \mid \xi\in F\}$ has a finite dimension and by $N_o(E)$ the set of $\mu\in N(E)$ for which the set*

$$\{\int\eta d\mu \mid \eta\in F \; , \; |\eta| \leq |\xi|\}$$

is compact for any $\xi\in F$.

Then:

a) for any continuous linear form φ on $(N_o(E),F)$ there exist $\xi\in F$, an equicontinuous $\sigma(E',E)$-closed set A' of E', and a real valued Radon measure λ on the compact space (Proposition 3.1.8)

$$\{\eta\in F \mid \; |\eta|\leq|\xi|\}\times A' ,$$

where $\{\eta\in F \mid |\eta| \leqslant |\xi|\}$ *is endowed with the topology induced by* $\sigma(F,N)$ *and* A' *by the topology induced by* $\sigma(E',E)$, *such that*

$$\varphi(\mu) = \int(\int \eta d(x'\circ\mu))d\lambda(\eta,x') ;$$

b) *if* E *possesses the approximation property then* $N_o(E)$ *is the closure of* $N_f(E)$ *in* $(N(E),F)$ *and* $(N_o(E),F)$ *possesses the approximation property.*

a) There exist $\xi\in F$ and an equicontinuous subset A' of E' such that $|\varphi(\mu)| \leqslant 1$ for any $\mu\in N_o(E)$ with $q_{\xi,A'}(\mu) \leqslant 1$. Let $\mu\in N_o(E)$. We set

$$K := \{\eta\in F \mid |\eta| \leqslant |\xi|\}$$

$$K_\mu := \{\int\eta d\mu \mid \eta\in F,\ |\eta| \leqslant |\xi|\} ,$$

and consider the maps

$$u_\mu : K \longrightarrow K_\mu ,\ \eta \longmapsto \int\eta d\mu ,$$

$$v_\mu : K_\mu\times A' \longrightarrow \mathbb{R} ,\ (x,x') \longmapsto \langle x,x'\rangle .$$

K_μ being compact the topologies induced on it by E and the weak topology of E coincide and therefore u_μ is continuous with respect to the topology induced by E on K . v_μ being continuous too the map $v_\mu\circ u_\mu$ is continuous. Let C be the Banach space of continuous real functions on $K\times A'$ (with the supremum norm) and let $\mathcal{D}$ be the subspace of C of the maps of the form $v_\mu\circ u_\mu$, where μ runs through $N_o(E)$. There exists a continuous linear form ψ on $\mathcal{D}$ such that

$$\psi(v_\mu\circ u_\mu) = \varphi(\mu)$$

for any $\mu\in N_o(E)$. By Hahn-Banach theorem we may even assume ψ is a continuous linear form on C . Let λ be a real valued Radon measure on $K\times A'$ such that

$$\int f d\lambda = \psi(f)$$

for any $f\in C$. We get

$$\varphi(\mu) = \psi(v_\mu \circ u_\mu) = \int(\int \eta d(x'\circ\mu))d\lambda(\eta,x') .$$

b) We show first $N_o(E)$ is closed. Let μ be an adherent point of $N_o(E)$ in $N(E)$, let U be a closed circled convex 0-neighbourhood in E, let U^o be the polar set of U in E', and let $\xi\in F$. There exists $\nu\in N_o(E)$ such that

$$\sup_{x'\in U^o} \int|\xi|d|x'\circ(\mu-\nu) \leq 1 .$$

By Proposition 4.3.6 b)

$$\int \eta d(\mu-\nu)\in U$$

for any $\eta\in F$ with $|\eta| \leq |\xi|$. Let M be a finite subset of E such that

$$\{\int\eta d\nu \,|\, \eta\in F,\ |\eta| \leq |\xi|\}\subset M+U .$$

We get

$$\{\int\eta d\mu \,|\, \eta\in F,\ |\eta| \leq |\xi|\}\subset M+2U .$$

Hence the set

$$\{\int\eta d\mu \,|\, \eta\in F,\ |\eta| \leq |\xi|\}$$

is precompact. Since it is weakly compact (Theorem 4.2.20 b)) it is compact.

Let now $\mu\in N_o(E)$, let A' be an equicontinuous subset of E', and let $\xi\in F$. We set

$$K := \{\int\eta d\mu \,|\, \eta\in F,\ |\eta| \leq |\xi|\} .$$

Since K is compact and since E possesses the approximation property there exists a finite family $((x_\iota,x'_\iota))_{\iota\in I}$ in $E\times E'$ such that

$$|\langle x,x'\rangle - \sum_{\iota\in I} \langle x,x'_\iota\rangle \langle x_\iota,x'\rangle| \leq 1$$

for any $(x,x')\in K\times A'$. We get

$$|\langle \int \eta d\mu - \int \eta d(\sum_{\iota\in I} (x'_\iota \circ \mu) x_\iota) , x' \rangle| \leqslant 1$$

for any $\eta\in F$ with $|\eta| \leqslant |\xi|$ and for any $x'\in A'$. This shows that μ is an adherent point of $N_f(E)$.

By Theorem 3.9.6 (N,F) possesses the approximation property. By Lemma 4.5.4 $(N_o(E),F)$ possesses the approximation property. □

Remark. The vector space $N_f(E)$ can be identified with $N\otimes E$.

Bibliography

[1] R.A. Alò, Approximation Integration, J. Math. Anal. Appl. 48 (1974), 127-138.

[2] R.G. Bartle, N. Dunford, J. Schwartz, Weak Compactness and Vector Measures, Canad. J. Math., 7 (1955), 289-305.

[3] I. Chiţescu, The Conjugate Space of the Space of Measures, Rev. Roumaine Math. Pures Appl., 21 (1976), 1313-1316.

[4] C. Constantinescu, Weakly Compact Sets in Locally Convex Vector Lattices, Rev. Roumaine Math, Pures Appl. 14 (1969), 325-351.

[5] C. Constantinescu, On Vector Measures, Ann. Inst. Fourier, 25, 3-4, (1975), 139-161.

[6] C. Constantinescu, Le dual de l'espace des measures sur un clan, C.R. Acad. Sci. Paris, 285 (1977), A47-A50.

[7] C. Constantinescu, Duality in Measure Theory, Preprint at the Technische Universität Hannover, Institut für Mathematik, Nr. 67 (1977).

[8] J. Diestel, J.J. Uhl Jr., Vector Measures, Amer. Math. Soc., Mathematical Surveys 15 (1977).

[9] J. Dixmier, Sur certains espaces considérés par M.H. Stone, Summa Brasiliensis Math. 2 (1951), 151-182.

[10] W.A. Feldman and J.F. Porter, Compact Convergence and the Order Bidual for C(X), Pacific J. Math., 57 (1975), 113-124.

[11] H. Gordon, The Maximal Ideal Space of a Ring of Measurable Functions, Amer. J. Math., 88 (1966), 827-843.

[12] W. Graves, D. Sentilles, The Extension and Completion of the Universal Measure and the Dual of the Space of Measures, J. Math. Anal. Appl., 68 (1979), 228-264.

[13] H.F. de Groote, Zur Integrationstheorie über bewerteten Körpern und der Darstellung des Biduals von $C_o(X,C)$, Manuscripta Math., 10 (1973), 65-89.

[14] A. Grothendieck, Sur les applications linéaires faiblement compactes d'espaces du type C(K), Canad. J. Math., 5 (1953), 129-173.

[15] A. Grothendieck, Produits tensoriels topologiques et espaces nucléaires, Mem. Amer. Math. Soc., 16 (1955).

[16] S. Kaplan, On the Second Dual of the Space of Continuous Functions I, II, III, IV, Trans. Amer. Math. Soc., I 86 (1957), 70-90; II 93 (1959), 329-350; III 101 (1961), 34-51; IV 113 (1964), 512-546.

[17] S. Kaplan, Closure Properties of C(X) in its Second Dual, Proc. Amer. Math, Soc. 17 (1966), 401-406.

[18] S. Kaplan, The Second Dual of the Space of Continuous Functions and the Riemann Integral, Illinois J. Math., 12 (1968), 283-302.

[19] S. Kaplan, The Unbounded Bidual of C(X), Illinois J. Math., 17 (1973), 619-645.

[20] A. Katsaras, D.B. Lin, Integral Representations of Weakly Compact Operators, Pacific J. Math., 56 (1975), 547-556.

[21] D.R. Lewis, Integration with Respect to Vector Measures, Pacific J. Math., 33 (1970), 157-165.

[22] W.A.J. Luxemburg and J.J. Masterson, An Extension of the Concept of the Order Dual of a Riesz Space, Canad. J. Math., 19 (1967) 488-498.

[23] J. Mack, The Order Dual of the Space of Radon Measures, Trans. Amer. Math. Soc. 113 (1964), 219-239.

[24] R.D. Mauldin, A Representation Theorem for the Second Dual of C [0,1], Studia Math., 44 (1973), 197-200

[25] R.D. Mauldin, The Continuum Hypothesis, Integration and Duals of Measure Spaces, Illinois J. Math., 19 (1975), 33-40.

[26] H.H. Schaefer, Topological Vector Spaces, Springer-Verlag, New York-Heidelberg-Berlin, (1971).

[27] Z. Semadeni, Banach Spaces of Continuous Functions I, Polish Scientific Publishers, Warszawa (1971).

[28] Ch. Shannon, The Second Dual of C(X), Pacific J. Math., 72 (1977), 237-253.

[29] Yu.A. Šreider, The Structure of Maximal Ideals in Rings of Measures with Convolution, Math. Sb., 27 (1950), 297-318 or Amer. Math. Soc. Trans., 81 (1953), 1-28.

[30] E. Thomas, L'intégration par rapport à une mesure de Radon vectorielle, Ann. Inst. Fourier, 20 (1970), 55-191.

[31] J. Tweddle, Weak Compactness in Locally Convex Spaces, Glasgow Math. J. 9 (1968), 123-127.

[32] B.Z. Vulikh, Introduction to the Theory of Partially Ordered Spaces, Wolters-Noordhoff Scientific Publications Ltd., Groningen (1967).

[33] J. Wada, Stonian Spaces and the Second Conjugate Spaces of AM-Spaces, Osaka Math.J. 9 (1957), 195-200.

Index

absolutely continuous 7

almost everywhere 7

approximation property 111

associated representations 32

Banach lattice 6

band 5

bounded filter 114

bounded measure 7

bounded representation 24

concassage 10

evaluation map 5

fundamental solid subspace 5

integrable 8,45,11

lattice norm 6

lattice seminorm 6

L-norm 6

L-seminorm 6

locally convex vector lattice 6

locally countable 9

locally μ-integrable 45, 119

measurable function 9

measurable set 9

measure 6

metric approximation property 108

M-norm 6

M-seminorm 6

N-concassage 10

normed vector lattice 6

null set 7

order complete vector lattice 5

order continuous 5, 116

order topology 106

positive linear form 5

pseudo-compact 105

quasi-σ-ring 121

real measure 6

representation 31

representation of μ associated to 26

ring of sets 6

solid 5

step function 7

strong approximation property 111

strong D.-P.-property 105

δ-filter 114

δ-quasi complete 114

δ-ring 6

μ-almost everywhere 7

μ-concassage 10

μ-integrable 8,45,119

μ-nullsset 7

Σ-complete 114

Σ-sequence 114

Notations

$C(Y)$ 17

$C_b(Y)$ ... 17

$C_c(Y)$ ... 17

$C_i(Y)$ ... 17

$C_\infty(Y)$ ... 16

E' 114

E'' 114

E^* 114

E_+ 5

E^+ 5

E^π 5

E^ρ 6

$\dot{f}$ 12

$f_1 \tilde{\times} f_2$ 92

$f|Y'$ 17

$f \cdot \mu$ 9,119

$(F,\mathcal{F})$ ... 178

F_G 114

$\tilde{\mathcal{G}}$ 64

$(\mathcal{G},\mathcal{F})$ ... 64

$\mathcal{L}$ 13

$\mathcal{L}_b$ 13

$\mathcal{L}_c$ 13

$\mathcal{L}_d$ 84

$\mathcal{L}_\infty$ 13

$\hat{\mathcal{L}}(\mu)$ 45

$\mathcal{L}^1(\mu)$ 8

$\hat{\mathcal{L}}^1(\mu)$ 45, 119

$\hat{\mathcal{L}}^p(\mu)$ 53

$\hat{L}^p(\mu)$ 53

$\hat{\mathcal{L}}^\perp(\mu)$ 45

$\mathcal{L}^1_{loc}(\mu)$ 9

$\hat{\mathcal{L}}^1_{loc}(\lambda,E)$ 146

$\hat{\mathcal{L}}^1_{loc}(\mu)$ 45,119

$\hat{\mathcal{L}}^p_{loc}(\mu)$ 53

$\hat{L}^1_{loc}(\lambda,E)$ 146

$\hat{L}^p_{loc}(\mu)$ 53

$\hat{L}^\infty(\mu)$ 90

$\mathcal{M}$ 12

$\mathcal{M}_b$ 12

$\mathcal{M}_c$ 12

$\mathcal{M}(f)$ 12

$\mathcal{M}(\underline{\underline{R}})$ 6

$\mathcal{M}(Y)$ 9

$\mathcal{M}(\xi)$ 45

$\mathcal{M}_1$ 91

$\mathcal{M}_2$ 91

$\mathcal{N}(F)$ 117

$(\mathcal{N},\mathcal{F})$ 64

$\mathcal{N}_1 \otimes \mathcal{N}_2$ 92

$\mathcal{N}_1 \bar{\otimes} \mathcal{N}_2$ 92

$(\mathcal{N}_1,\mathcal{F}_1) \otimes (\mathcal{N}_2,\mathcal{F}_2)$ ... 92

$(\mathcal{N}_1,\mathcal{F}_1) \bar{\bar{\otimes}} (\mathcal{N}_2,\mathcal{F}_2)$ ... 92

p_ξ 64

q_ξ 64

$q_{\xi,A'}$ 177

$\underline{\underline{R}}$ 7, 12

$\underline{\underline{R}}_1$ 91

$\underline{\underline{R}}_2$ 91

$\underline{\underline{R}}_1 \otimes \underline{\underline{R}}_2$ 91

r_ξ 84

$\underline{\underline{S}}$ 91

X 7, 12

X_1 91

X_2 91

x_+ 5

x_- 5

$|x|$ 5

μ-a.e. 7

$\mu|Y'$ 17

$\mu_1 \tilde{\times} \mu_2$ 91

$|\xi|^p$ 53

ξ_η 49

$\xi\varphi$ 147

$\xi\cdot\mu$ 45, 119

$\xi_1 \tilde{\times} \xi_2$ 94

$\hat{\varphi}$ 47

$(\varphi\cdot\lambda)(A)$ 146

1_A 7

1_A^X 7, 91

$\int f d\mu$ 8, 119

$\int_A f d\mu$ 8

$\int \xi d\mu$ 45, 119

$\| \ \|$ 7, 89

$\vee$ 5

$\wedge$ 5

$\overset{E}{\vee}$ 5

$\overset{E}{\wedge}$ 5

Vol. 640: J. L. Dupont, Curvature and Characteristic Classes. X, 175 pages. 1978.

Vol. 641: Séminaire d'Algèbre Paul Dubreil, Proceedings Paris 1976–1977. Edité par M. P. Malliavin. IV, 367 pages. 1978.

Vol. 642: Theory and Applications of Graphs, Proceedings, Michigan 1976. Edited by Y. Alavi and D. R. Lick. XIV, 635 pages. 1978.

Vol. 643: M. Davis, Multiaxial Actions on Manifolds. VI, 141 pages. 1978.

Vol. 644: Vector Space Measures and Applications I, Proceedings 1977. Edited by R. M. Aron and S. Dineen. VIII, 451 pages. 1978.

Vol. 645: Vector Space Measures and Applications II, Proceedings 1977. Edited by R. M. Aron and S. Dineen. VIII, 218 pages. 1978.

Vol. 646: O. Tammi, Extremum Problems for Bounded Univalent Functions. VIII, 313 pages. 1978.

Vol. 647: L. J. Ratliff, Jr., Chain Conjectures in Ring Theory. VIII, 133 pages. 1978.

Vol. 648: Nonlinear Partial Differential Equations and Applications, Proceedings, Indiana 1976–1977. Edited by J. M. Chadam. VI, 206 pages. 1978.

Vol. 649: Séminaire de Probabilités XII, Proceedings, Strasbourg, 1976–1977. Edité par C. Dellacherie, P. A. Meyer et M. Weil. VIII, 805 pages. 1978.

Vol. 650: C*-Algebras and Applications to Physics. Proceedings 1977. Edited by H. Araki and R. V. Kadison. V, 192 pages. 1978.

Vol. 651: P. W. Michor, Functors and Categories of Banach Spaces. VI, 99 pages. 1978.

Vol. 652: Differential Topology, Foliations and Gelfand-Fuks-Cohomology, Proceedings 1976. Edited by P. A. Schweitzer. XIV, 252 pages. 1978.

Vol. 653: Locally Interacting Systems and Their Application in Biology. Proceedings, 1976. Edited by R. L. Dobrushin, V. I. Kryukov and A. L. Toom. XI, 202 pages. 1978.

Vol. 654: J. P. Buhler, Icosahedral Golois Representations. III, 143 pages. 1978.

Vol. 655: R. Baeza, Quadratic Forms Over Semilocal Rings. VI, 199 pages. 1978.

Vol. 656: Probability Theory on Vector Spaces. Proceedings, 1977. Edited by A. Weron. VIII, 274 pages. 1978.

Vol. 657: Geometric Applications of Homotopy Theory I, Proceedings 1977. Edited by M. G. Barratt and M. E. Mahowald. VIII, 459 pages. 1978.

Vol. 658: Geometric Applications of Homotopy Theory II, Proceedings 1977. Edited by M. G. Barratt and M. E. Mahowald. VIII, 487 pages. 1978.

Vol. 659: Bruckner, Differentiation of Real Functions. X, 247 pages. 1978.

Vol. 660: Equations aux Dérivée Partielles. Proceedings, 1977. Edité par Pham The Lai. VI, 216 pages. 1978.

Vol. 661: P. T. Johnstone, R. Paré, R. D. Rosebrugh, D. Schumacher, R. J. Wood, and G. C. Wraith, Indexed Categories and Their Applications. VII, 260 pages. 1978.

Vol. 662: Akin, The Metric Theory of Banach Manifolds. XIX, 306 pages. 1978.

Vol. 663: J. F. Berglund, H. D. Junghenn, P. Milnes, Compact Right Topological Semigroups and Generalizations of Almost Periodicity. X, 243 pages. 1978.

Vol. 664: Algebraic and Geometric Topology, Proceedings, 1977. Edited by K. C. Millett. XI, 240 pages. 1978.

Vol. 665: Journées d'Analyse Non Linéaire. Proceedings, 1977. Edité par P. Bénilan et J. Robert. VIII, 256 pages. 1978.

Vol. 666: B. Beauzamy, Espaces d'Interpolation Réels: Topologie et Géometrie. X, 104 pages. 1978.

Vol. 667: J. Gilewicz, Approximants de Padé. XIV, 511 pages. 1978.

Vol. 668: The Structure of Attractors in Dynamical Systems. Proceedings, 1977. Edited by J. C. Martin, N. G. Markley and W. Perrizo. VI, 264 pages. 1978.

Vol. 669: Higher Set Theory. Proceedings, 1977. Edited by G. H. Müller and D. S. Scott. XII, 476 pages. 1978.

Vol. 670: Fonctions de Plusieurs Variables Complexes III, Proceedings, 1977. Edité par F. Norguet. XII, 394 pages. 1978.

Vol. 671: R. T. Smythe and J. C. Wierman, First-Passage Perculation on the Square Lattice. VIII, 196 pages. 1978.

Vol. 672: R. L. Taylor, Stochastic Convergence of Weighted Sums of Random Elements in Linear Spaces. VII, 216 pages. 1978.

Vol. 673: Algebraic Topology, Proceedings 1977. Edited by P. Hoffman, R. Piccinini and D. Sjerve. VI, 278 pages. 1978.

Vol. 674: Z. Fiedorowicz and S. Priddy, Homology of Classical Groups Over Finite Fields and Their Associated Infinite Loop Spaces. VI, 434 pages. 1978.

Vol. 675: J. Galambos and S. Kotz, Characterizations of Probability Distributions. VIII, 169 pages. 1978.

Vol. 676: Differential Geometrical Methods in Mathematical Physics II, Proceedings, 1977. Edited by K. Bleuler, H. R. Petry and A. Reetz. VI, 626 pages. 1978.

Vol. 677: Séminaire Bourbaki, vol. 1976/77, Exposés 489–506. IV, 264 pages. 1978.

Vol. 678: D. Dacunha-Castelle, H. Heyer et B. Roynette. Ecole d'Eté de Probabilités de Saint-Flour. VII-1977. Edité par P. L. Hennequin. IX, 379 pages. 1978.

Vol. 679: Numerical Treatment of Differential Equations in Applications, Proceedings, 1977. Edited by R. Ansorge and W. Törnig. IX, 163 pages. 1978.

Vol. 680: Mathematical Control Theory, Proceedings, 1977. Edited by W. A. Coppel. IX, 257 pages. 1978.

Vol. 681: Séminaire de Théorie du Potentiel Paris, No. 3, Directeurs: M. Brelot, G. Choquet et J. Deny. Rédacteurs: F. Hirsch et G. Mokobodzki. VII, 294 pages. 1978.

Vol. 682: G. D. James, The Representation Theory of the Symmetric Groups. V, 156 pages. 1978.

Vol. 683: Variétés Analytiques Compactes, Proceedings, 1977. Edité par Y. Hervier et A. Hirschowitz. V, 248 pages. 1978.

Vol. 684: E. E. Rosinger, Distributions and Nonlinear Partial Differential Equations. XI, 146 pages. 1978.

Vol. 685: Knot Theory, Proceedings, 1977. Edited by J. C. Hausmann. VII, 311 pages. 1978.

Vol. 686: Combinatorial Mathematics, Proceedings, 1977. Edited by D. A. Holton and J. Seberry. IX, 353 pages. 1978.

Vol. 687: Algebraic Geometry, Proceedings, 1977. Edited by L. D. Olson. V, 244 pages. 1978.

Vol. 688: J. Dydak and J. Segal, Shape Theory. VI, 150 pages. 1978.

Vol. 689: Cabal Seminar 76–77, Proceedings, 1976–77. Edited by A.S. Kechris and Y. N. Moschovakis. V, 282 pages. 1978.

Vol. 690: W. J. J. Rey, Robust Statistical Methods. VI, 128 pages. 1978.

Vol. 691: G. Viennot, Algèbres de Lie Libres et Monoïdes Libres. III, 124 pages. 1978.

Vol. 692: T. Husain and S. M. Khaleelulla, Barrelledness in Topological and Ordered Vector Spaces. IX, 258 pages. 1978.

Vol. 693: Hilbert Space Operators, Proceedings, 1977. Edited by J. M. Bachar Jr. and D. W. Hadwin. VIII, 184 pages. 1978.

Vol. 694: Séminaire Pierre Lelong – Henri Skoda (Analyse) Année 1976/77. VII, 334 pages. 1978.

Vol. 695: Measure Theory Applications to Stochastic Analysis, Proceedings, 1977. Edited by G. Kallianpur and D. Kölzow. XII, 261 pages. 1978.

Vol. 696: P. J. Feinsilver, Special Functions, Probability Semigroups, and Hamiltonian Flows. VI, 112 pages. 1978.

Vol. 697: Topics in Algebra, Proceedings, 1978. Edited by M. F. Newman. XI, 229 pages. 1978.

Vol. 698: E. Grosswald, Bessel Polynomials. XIV, 182 pages. 1978.

Vol. 699: R. E. Greene and H.-H. Wu, Function Theory on Manifolds Which Possess a Pole. III, 215 pages. 1979.

Vol. 700: Module Theory, Proceedings, 1977. Edited by C. Faith and S. Wiegand. X, 239 pages. 1979.

Vol. 701: Functional Analysis Methods in Numerical Analysis, Proceedings, 1977. Edited by M. Zuhair Nashed. VII, 333 pages. 1979.

Vol. 702: Yuri N. Bibikov, Local Theory of Nonlinear Analytic Ordinary Differential Equations. IX, 147 pages. 1979.

Vol. 703: Equadiff IV, Proceedings, 1977. Edited by J. Fábera. XIX, 441 pages. 1979.

Vol. 704: Computing Methods in Applied Sciences and Engineering, 1977, I. Proceedings, 1977. Edited by R. Glowinski and J. L. Lions. VI, 391 pages. 1979.

Vol. 705: O. Forster und K. Knorr, Konstruktion verseller Familien kompakter komplexer Räume. VII, 141 Seiten. 1979.

Vol. 706: Probability Measures on Groups, Proceedings, 1978. Edited by H. Heyer. XIII, 348 pages. 1979.

Vol. 707: R. Zielke, Discontinuous Čebyšev Systems. VI, 111 pages. 1979.

Vol. 708: J. P. Jouanolou, Equations de Pfaff algébriques. V, 255 pages. 1979.

Vol. 709: Probability in Banach Spaces II. Proceedings, 1978. Edited by A. Beck. V, 205 pages. 1979.

Vol. 710: Séminaire Bourbaki vol. 1977/78, Exposés 507–524. IV, 328 pages. 1979.

Vol. 711: Asymptotic Analysis. Edited by F. Verhulst. V, 240 pages. 1979.

Vol. 712: Equations Différentielles et Systèmes de Pfaff dans le Champ Complexe. Edité par R. Gérard et J.-P. Ramis. V, 364 pages. 1979.

Vol. 713: Séminaire de Théorie du Potentiel, Paris No. 4. Edité par F. Hirsch et G. Mokobodzki. VII, 281 pages. 1979.

Vol. 714: J. Jacod, Calcul Stochastique et Problèmes de Martingales. X, 539 pages. 1979.

Vol. 715: Inder Bir S. Passi, Group Rings and Their Augmentation Ideals. VI, 137 pages. 1979.

Vol. 716: M. A. Scheunert, The Theory of Lie Superalgebras. X, 271 pages. 1979.

Vol. 717: Grosser, Bidualräume und Vervollständigungen von Banachmoduln. III, 209 pages. 1979.

Vol. 718: J. Ferrante and C. W. Rackoff, The Computational Complexity of Logical Theories. X, 243 pages. 1979.

Vol. 719: Categorial Topology, Proceedings, 1978. Edited by H. Herrlich and G. Preuß. XII, 420 pages. 1979.

Vol. 720: E. Dubinsky, The Structure of Nuclear Fréchet Spaces. V, 187 pages. 1979.

Vol. 721: Séminaire de Probabilités XIII. Proceedings, Strasbourg, 1977/78. Edité par C. Dellacherie, P. A. Meyer et M. Weil. VII, 647 pages. 1979.

Vol. 722: Topology of Low-Dimensional Manifolds. Proceedings, 1977. Edited by R. Fenn. VI, 154 pages. 1979.

Vol. 723: W. Brandal, Commutative Rings whose Finitely Generated Modules Decompose. II, 116 pages. 1979.

Vol. 724: D. Griffeath, Additive and Cancellative Interacting Particle Systems. V, 108 pages. 1979.

Vol. 725: Algèbres d'Opérateurs. Proceedings, 1978. Edité par P. de la Harpe. VII, 309 pages. 1979.

Vol. 726: Y.-C. Wong, Schwartz Spaces, Nuclear Spaces and Tensor Products. VI, 418 pages. 1979.

Vol. 727: Y. Saito, Spectral Representations for Schrödinger Operators With Long-Range Potentials. V, 149 pages. 1979.

Vol. 728: Non-Commutative Harmonic Analysis. Proceedings, 1978. Edited by J. Carmona and M. Vergne. V, 244 pages. 1979.

Vol. 729: Ergodic Theory. Proceedings, 1978. Edited by M. Denker and K. Jacobs. XII, 209 pages. 1979.

Vol. 730: Functional Differential Equations and Approximation of Fixed Points. Proceedings, 1978. Edited by H.-O. Peitgen and H.-O. Walther. XV, 503 pages. 1979.

Vol. 731: Y. Nakagami and M. Takesaki, Duality for Crossed Products of von Neumann Algebras. IX, 139 pages. 1979.

Vol. 732: Algebraic Geometry. Proceedings, 1978. Edited by K. Lønsted. IV, 658 pages. 1979.

Vol. 733: F. Bloom, Modern Differential Geometric Techniques in the Theory of Continuous Distributions of Dislocations. XII, 206 pages. 1979.

Vol. 734: Ring Theory, Waterloo, 1978. Proceedings, 1978. Edited by D. Handelman and J. Lawrence. XI, 352 pages. 1979.

Vol. 735: B. Aupetit, Propriétés Spectrales des Algèbres de Banach. XII, 192 pages. 1979.

Vol. 736: E. Behrends, M-Structure and the Banach-Stone Theorem. X, 217 pages. 1979.

Vol. 737: Volterra Equations. Proceedings 1978. Edited by S.-O. Londen and O. J. Staffans. VIII, 314 pages. 1979.

Vol. 738: P. E. Conner, Differentiable Periodic Maps. 2nd edition, IV, 181 pages. 1979.

Vol. 739: Analyse Harmonique sur les Groupes de Lie II. Proceedings, 1976–78. Edited by P. Eymard et al. VI, 646 pages. 1979.

Vol. 740: Séminaire d'Algèbre Paul Dubreil. Proceedings, 1977–78. Edited by M.-P. Malliavin. V, 456 pages. 1979.

Vol. 741: Algebraic Topology, Waterloo 1978. Proceedings. Edited by P. Hoffman and V. Snaith. XI, 655 pages. 1979.

Vol. 742: K. Clancey, Seminormal Operators. VII, 125 pages. 1979.

Vol. 743: Romanian-Finnish Seminar on Complex Analysis. Proceedings, 1976. Edited by C. Andreian Cazacu et al. XVI, 713 pages. 1979.

Vol. 744: I. Reiner and K. W. Roggenkamp, Integral Representations. VIII, 275 pages. 1979.

Vol. 745: D. K. Haley, Equational Compactness in Rings. III, 167 pages. 1979.

Vol. 746: P. Hoffman, τ-Rings and Wreath Product Representations. V, 148 pages. 1979.

Vol. 747: Complex Analysis, Joensuu 1978. Proceedings, 1978. Edited by I. Laine, O. Lehto and T. Sorvali. XV, 450 pages. 1979.

Vol. 748: Combinatorial Mathematics VI. Proceedings, 1978. Edited by A. F. Horadam and W. D. Wallis. IX, 206 pages. 1979.

Vol. 749: V. Girault and P.-A. Raviart, Finite Element Approximation of the Navier-Stokes Equations. VII, 200 pages. 1979.

Vol. 750: J. C. Jantzen, Moduln mit einem höchsten Gewicht. III, 195 Seiten. 1979.

Vol. 751: Number Theory, Carbondale 1979. Proceedings. Edited by M. B. Nathanson. V, 342 pages. 1979.

Vol. 752: M. Barr, *-Autonomous Categories. VI, 140 pages. 1979.

Vol. 753: Applications of Sheaves. Proceedings, 1977. Edited by M. Fourman, C. Mulvey and D. Scott. XIV, 779 pages. 1979.

Vol. 754: O. A. Laudal, Formal Moduli of Algebraic Structures. III, 161 pages. 1979.

Vol. 755: Global Analysis. Proceedings, 1978. Edited by M. Grmela and J. E. Marsden. VII, 377 pages. 1979.

Vol. 756: H. O. Cordes, Elliptic Pseudo-Differential Operators – An Abstract Theory. IX, 331 pages. 1979.

Vol. 757: Smoothing Techniques for Curve Estimation. Proceedings, 1979. Edited by Th. Gasser and M. Rosenblatt. V, 245 pages. 1979.

Vol. 758: C. Năstăsescu and F. Van Oystaeyen; Graded and Filtered Rings and Modules. X, 148 pages. 1979.

This series reports new developments in mathematical research and teaching – quickly, informally and at a high level. The type of material considered for publication includes:

1. Preliminary drafts of original papers and monographs
2. Lectures on a new field or presentations of a new angle in a classical field
3. Seminar work-outs
4. Reports of meetings, provided they are
   a) of exceptional interest and
   b) devoted to a single topic.

Texts which are out of print but still in demand may also be considered if they fall within these categories.

The timeliness of a manuscript is more important than its form, which may be unfinished or tentative. Thus, in some instances, proofs may be merely outlined and results presented which have been or will later be published elsewhere. If possible, a subject index should be included. Publication of Lecture Notes is intended as a service to the international mathematical community, in that a commercial publisher, Springer-Verlag, can offer a wide distribution of documents which would otherwise have a restricted readership. Once published and copyrighted, they can be documented in the scientific literature.

Manuscripts

Manuscripts should be no less than 100 and preferably no more than 500 pages in length.
They are reproduced by a photographic process and therefore must be typed with extreme care. Symbols not on the typewriter should be inserted by hand in indelible black ink. Corrections to the typescript should be made by pasting in the new text or painting out errors with white correction fluid. Authors receive 75 free copies and are free to use the material in other publications. The typescript is reduced slightly in size during reproduction; best results will not be obtained unless the text on any one page is kept within the overall limit of 18 x 26.5 cm (7 x 10½ inches). On request, the publisher will supply special paper with the typing area outlined.

Manuscripts should be sent to Prof. A. Dold, Mathematisches Institut der Universität Heidelberg, Im Neuenheimer Feld 288, 6900 Heidelberg/Germany, Prof. B. Eckmann, Eidgenössische Technische Hochschule, CH-8092 Zürich/Switzerland, or directly to Springer-Verlag Heidelberg.

Springer-Verlag, Heidelberger Platz 3, D-1000 Berlin 33
Springer-Verlag, Neuenheimer Landstraße 28–30, D-6900 Heidelberg 1
Springer-Verlag, 175 Fifth Avenue, New York, NY 10010/USA

ISBN 3-540-**09989-1**
ISBN 0-387-**09989-1**